D0241649

The Odd Body

Dr Stephen Juan

The Odd Body

illustrated by Olivier Kugler

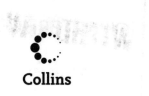

Collins

First published in Australia in 1995
This edition published in 2004
by Collins, an imprint of HarperCollins Publishers

HarperCollins Publishers
77-85 Fulham Palace Road
London w6 8jb
www.collins.co.uk

A cip catalogue record for this book is available
from the British Library

isbn 0-00-718521-9

Contents

Acknowledgements

There are many people to thank for making this book possible. At the top of the list are the scientists and researchers who investigate the human body and how it works. Some are from the medical sciences, some from the behavioural sciences, and some from elsewhere. This book would have nothing to say if there was no body of knowledge to communicate. Next are the curious people who ask questions, especially those who can't sit still until they get the answers. Many of these are my own university students who have asked me questions about 'us', which I have always tried to answer as best I could. Curiosity may have killed the cat, but it also leads to knowledge. Thus it is always to be encouraged. And thanks to those individuals who have sparked my own curiosity from what they have written or said.

Thanks are also due to the librarians at the University of Sydney who make resources available and who are always co-operative and helpful.

Thanks too to the people at HarperCollins Australia who encouraged me to undertake this project: Linda Tenkate, Melissa Gabbott, Mel Cox, Jude McGee, Lisa Mills and Angelo Loukakis.

Thanks finally to my wife, Buffy, and to our two daughters, Alicia and Cassie, who gave me time and space to produce whatever of value is between these pages.

Introduction

¶ Has this ever happened to you?

Did you ever want to know something about the human body but were afraid to ask? Or you didn't know who to ask? Or there was nobody around to ask? Say you wanted to know about why you yawn, why your skin wrinkles after a bath, something silly like why do men have nipples, or something really weird like can you keep a severed head alive? At home you may have thought about asking your parents. Maybe you even tried. But more often than not they didn't know themselves. If they told you to 'look it up' (the face-saving suggestion to maintain a parent's dignity when faced with their own ignorance) and you did, you probably couldn't find a book that gave you the answer you wanted. So you put the question to the back of your mind and eventually forgot about it. A few years later, in biology/life education classes in school, the question may have occurred to you again. Should you ask the teacher? You decided not to. After all, the question was off the subject, it would take up class time, your friends might think you were 'weird', Mr Fletcher probably didn't know anyway and, after all, it wouldn't be on the exam. So you put off your question again and eventually forgot it again.

Now you're an adult. You're in your doctor's office for your annual check-up. No particular problems, but out of the blue you remember that question you first asked yourself when you were a kid. Should you ask the doctor? After all, doctors are trained in this sort of thing. They ought to know everything about the body since it's their job to fix it when it's broken. But you hesitate. The doctor is busy. There are other patients waiting. And, after all, your question doesn't relate to your health or to any illness you're likely to get. So you put off your question yet again and forget it again.

Has this ever happened to you?

If so, then this book is for you. You can stop putting off your questions about the human body. Chances are the answer is here. *The Odd Body* tries to explain all of those body mysteries you've had, both major and minor, whether for a short or long time. We call these OBQs – odd body questions. We ourselves have been asking these sorts of questions for many years – more than we'd like to admit. We love the commonplace questions, the silly, the weird, the bizarre, the fascinating. We hope your question is answered here. Perhaps too there are a few questions within these pages that you never thought to ask. It might be fun to find out about them just the same.

If there's any real lesson in this book it's simply this: human beings are so interesting. Finding out more about us is one of the true pleasures of life.

The Odd Body

1 · Beginnings

Many of us ask questions about our origins, our in utero development, and how we are born into this world. It's said that we come into this world with nothing. But that's only the beginning of the story.

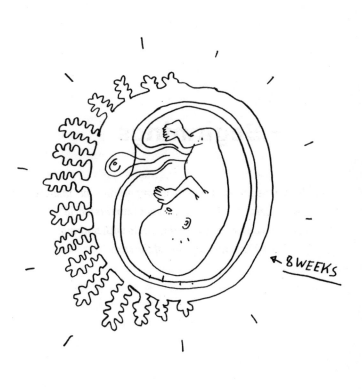

8 WEEKS

¶ What makes me a human being?

We are humans because we are classified as such based on our unique physical and cultural characteristics. We manipulate symbols, express ourselves through language, and possess an enormous capacity to develop the intricacies of culture.

Taxonomy is the science of classifying life forms. As science classifies humans, we are members of the animal kingdom, the metazoan sub-kingdom, the chordata phylum, the vertebrata sub-phylum, the class mammalia, the sub-class theria, the infra-class eutheria, and the primate order. After this, it starts to get extremely interesting.

The suborder called anthropoidea is within the primate order. This suborder includes monkeys and apes as well as humans. Within the anthropoidea is the superfamily called hominoidea. This superfamily includes the anthropoid apes and both extinct and modern humans. It excludes the non-anthropoid apes. Anthropoid apes are tailless and include the gibbon, chimpanzee, gorilla and orang-utan. Within the hominoids is the family called hominidea or hominids. Hominids include both modern and extinct forms of human beings. It excludes the anthropoid apes.

What makes the hominids so special is a large brain and the ability to walk on two legs (bi-pedalism). It is a line-ball decision as to where to draw the line between 'human' and our 'human-like' ancestors. One place to draw it is to simply include all hominids as humans.

As for the beginnings of the earliest hominids – the beginnings of us – anthropologists have been pushing the date back for most of this century as new fossil evidence is revealed.

In 1974, a female hominid skeleton, nearly 40 per cent complete, was found by Dr Donald Johanson and T. Gray of the Institute of Human Origins in Berkeley, California at a site near Hadar in Ethiopia. Nicknamed 'Lucy', she was estimated to have lived for 40 years and attained the height of 106 cm. Lucy was dated at 3.2 million years old.

In 1978, fossilised footprints and parallel tracks left in volcanic ash and extending over a distance of 24 m were discovered by Dr Mary Leakey and Paul Abell near Laetoli in Tanzania. The three obviously hominid beings that left behind the prints and tracks were estimated to be no shorter than 120 cm tall. The fossils were dated at 3.6 million years old.

In 1984, a hominid jawbone with two molars 5 cm long was found by Kiptalam Chepboi in the Lake Baringo region of Kenya. It has been dated at 4 million years old.

In 1994, Drs Johanson and William Kimbel, along with Dr Yoel Rak of the University of Tel Aviv, reported finding fragments of a hominid skull as well as a number of limbs and jawbones at Hadar. These were dated as being about the same age as Lucy but this hominid was much taller.[1]

Later in 1994, Drs Tim White from the Department of Anthropology at the University of California at Berkeley, Gen Suwa from the University of Tokyo, and Berhane Asfaw from the Ethiopian government reported finding part of a child's jaw and two teeth at a site near the village of Aramis, 65 km south of Hadar. These fossils have been dated at 4.4 million years old – the earliest hominid remains so far.

The existence of this last fossil supports the theory that a common hominoid ancestor for all hominids lived in Africa no more than 6 million years ago.[2]

Humans are also Homo sapiens. We belong to the genus of Homo and the species of sapiens.

The earliest member of the Homo genus is the Homo habilis or 'handy man'. In 1964, part of a 'handy man's' skull was found at Olduvai Gorge in Tanzania and named by Drs Louis Leakey, Philip Tobias and John Napier, with the assistance of Raymond Dart. The following year another skull fragment was found in western Kenya but not dated until 1991. The oldest 'handy man' remains have been dated at 2.4 million years old.

Homo erectus is the nearest direct ancestor to Homo sapiens. In 1985, Kamoya Kimeu found the earliest remains of Homo erectus at a site near Lake Turkana in Kenya. It was a nearly complete skeleton of a twelve-year-old boy who stood 165 cm tall. The skeleton was dated at 1.6 million years old.

The earliest human tools were found in 1976 by Drs Helene Roche and John Wall near Hadar. These basic stone implements used for chopping and slicing have been dated at 2.7 million years old.

¶ When did I first know I was alive?

We probably know that we are alive sometime before we are born but it is difficult to remember this. It is theorised that we fail to remember because we do not have language to hold on to the memory.

The foetus becomes conscious sometime during the second trimester of pregnancy. Tactile sensitivity begins as early as the seventh week when a foetus first reacts to the stroke of a hair on its cheek. Skin sensitivity expands to include most parts of the body by the seventeenth week.[3] From the sixteenth week onwards, the unborn baby is easily startled by loud noises and turns away when a bright light is flashed on its mother's abdomen. The foetus reacts actively to rock music by kicking frantically. Interestingly, the foetus reacts in the opposite manner to calm music. It is unlikely that anything other than the physical sensation of sound can be heard by a foetus. It is rather like the noise that one hears from a distant house where a stereo is blasting away. You can hear the pulsation of the bass, but cannot distinguish the lyrics.

Even as early as twelve weeks, the foetus can be observed apparently squinting and scowling. At 14 weeks, it apparently sneers and looks dissatisfied. But at 24 weeks, the foetus shows behaviour which may indicate true thinking (cognition). The foetus can be observed frowning, grimacing, and smiling. But more importantly, while being viewed via ultrasound, a foetus at 24 weeks who was accidentally hit by a needle during an amniocentesis, was observed twisting its body away, locating the needle with its arm, and repeatedly striking the barrel of the needle with its arm and hand.[4]

There is speculation that the foetus must be thinking when it demonstrates anxiety. At 24 weeks, the foetus seems to be anxious since it can be observed sucking its thumb – sometimes so hard that blisters are raised.

By 26 weeks, the foetus can do some rather interesting gymnastics inside the womb. For example, it can do an elegant forward roll. It is speculative whether such movements are intentional and thus indicative of thinking.[5]

¶ When did I first dream?

Other evidence of thinking concerns dreaming. There is evidence that the foetus dreams. Indeed, the foetus dreams more than newborn infants who, in turn, dream more than older children, who dream more than adults. Sonographic studies show that REM sleep (rapid eye movement sleep in which dreaming takes place) occurs at 23 weeks. This is virtually the only type of sleep the foetus engages in. It is only at 36 weeks that non- REM sleep is detected. Thus, one could almost say for a foetus after 23 weeks, whenever it is sleeping it is dreaming.[6]

¶ When did I first feel?

Solid evidence suggests that the foetus feels pain by no later than 26 weeks. Yet some claim this ability occurs much earlier. One study suggested that the foetus feels pain by the seventh week.[7]

Pain pathways in the brain, as well as the cortical and subcortical centres necessary for pain perception, are well developed by the third trimester. Responses to painful stimuli have been documented in newborns (neonates) of all viable gestational ages.

In 1969, Dr Davenport Hooker at the University of Pittsburgh found that a foetus aborted during the thirteenth week (but not yet dead) will respond reflexively to the touch of a hair around its mouth. He also reported that a baby born three months prematurely will respond reflexively to the touch of a hair anywhere on its body.[8]

There is every indication that a newborn baby is in some ways just as sensitive to touch as older humans are. A newborn's skin is thinner than an adult's. As such, its nerve endings are less well insulated. Moreover, a newborn's nerve endings are just as mature and far more numerous than an adult's. The portion of the brain that processes touch sensations (the somatosensory cortex) is more developed at birth than any other portion of the brain.[9]

Nevertheless, it takes years for the sense of touch

to develop fully. Children cannot distinguish most objects by touch alone until they are about six or seven years old. The first foetal touch receptors appear on the skin by no later than the tenth week – while still surrounded by water. Nevertheless, according to Dr Maria Fitzgerald, professor of developmental neurobiology at the University of London, 'although the foetus lives in fluid, it never feels wetness'.[10] It is just like a person swimming underwater not feeling the water as such, but 'will notice the pressure of the wave'.[11]

¶ When did I first see?

Vision develops to some extent before birth. However, the newborn is very nearsighted. The foetal eyelids form by 10 weeks but remain fused shut until the twenty-sixth week at the latest. Nevertheless, the foetus will react to lights flashed on the mother's abdomen.[12]

Visually, babies are fascinated by two things. These are the human face and high-contrast geometrics. Briefly, the general thrust of research in this area leads to the following conclusions:

‖ From birth to about two months of age, babies see objects best at close range. This is about 20 cm from the eyes at birth and about 30 cm away at six weeks. They can discriminate differences in shape, size, and pattern and are more attracted to high-contrast patterns than to colour or brightness alone. They prefer to look at patterns of simple to moderate complexity and look more frequently at outside edges than internal patterns.

‖ From about two months to four months of age, babies scan their entire vision field and explore both interior patterns and outside edges. They prefer patterns of increasing complexity and curved lines and shapes to straight lines or angular shapes. They are especially attracted to faces and shapes. Babies begin to show that they remember what they see.

|| After about four months of age, babies adjust their focus to see near or far objects. They see in full colour and continue to prefer curved patterns and shapes. They seek out complexity and novelty in their visual environment and begin to develop depth perception.[13]

Children usually learn to identify colours between the ages of three and seven. If they seriously confuse colours after this time, then colour blindness is a distinct possibility.

¶ When did I first hear?

The foetal listening system begins to function by 16 weeks, even before the ear is complete.[14]

Surprisingly perhaps, the sense of hearing in a foetus begins with the skin. According to Dr David Chamberlain, president of the Association for Pre- and Perinatal Psychology and Health in Arlington, Virginia, the skin is 'a multisensory receptor organ integrating input from mechanoreceptors, thermo receptors, and pain receptors (nocireceptors). This early form of hearing is linked with the vestibular system which is sensitive to gravity and space, and with the cochlear system as it forms.'[15]

The hearing of the newborn is excellent. We have known this for decades from experiments with the startle reflex. In fact, hearing is far more mature than vision in the newborn. In a series of classic experiments, it was demonstrated that before the infant is fully delivered from the birth canal, when just the head is popping through, if a sound is made at one side or the other of the head, the infant's eyes will turn toward the sound – as if the baby knows that something is there to be seen.[16] It is also interesting to note that the newborn hears as well when sleeping as when awake.

¶ When did I first smell?

Although the foetus is surrounded by fluid, it can definitely smell. However, according to Dr Stephen Roper, professor of anatomy and neurobiology at Colorado State University in Fort Collins, 'the foetus doesn't sniff. Rather, odours are absorbed by nasal tissues.'[17] Indeed, many species of fish have this same ability.

The amniotic fluid that the foetus swims in is full of odours. If the mother eats spicy foods, the fluid can smell like a Mediterranean salad. Also, subtle smells in the fluid are unique to the mother – just as body odour is unique. After birth, these smells may help solidify the baby-mother relationship.

Immediately after birth, the newborn cannot smell through its nose because it is clogged with amniotic fluid and other substances for about a day. This clogging effect resembles the stuffiness of an adult's nose. A baby's sense of smell starts to emerge right after the clogging stops – two days after birth. There is even evidence that babies who are just a few days old have as good a sense of smell as adults.

As early as 1934, Dr Dorothy Disher found that one-month-old babies were more likely to wriggle in their cribs when presented with a number of odours compared with smelling merely pure air. Babies responded most to the smells of violet, asafoetida, sassafras, citronella, turpentine, pyridine, and lemon.[18]

In a now classic laboratory experiment, Dr Jacob Steiner of the Hadassah School of Dentistry at the Hebrew University in Jerusalem asked a panel of adults to select the most 'fresh' and the most 'rotten' from among a large collection of odours. The adults judged that honey was the freshest followed by banana, vanilla, and chocolate smells. Rotting eggs followed by rotting prawns were unanimously judged the most rotten smells. Dr Steiner then held swabs of these odours under the noses of babies only a few hours old. They smiled when smelling the fresh odours and grimaced when smelling the rotten ones. Furthermore, the widest smiles came from smelling the honey and the biggest grimaces came from smelling the rotting eggs – the identical choices made by the adults.[19]

Other researchers have found evidence that a majority of newborn babies can smell better than adults – in this case, the researcher. In 1975, Oxford University psychologist, Dr Aidan Macfarlane, tested whether a newborn could tell the difference between the smell of its own mother (and her milk) compared with the smell of another baby's mother (and her milk). The smells came from gauze pads that the mothers had kept within their bras to absorb any milk leaking from their breasts.

Dr Macfarlane draped a pad from the baby's mother along one side of the baby's face. Along the other side, he draped a pad from another mother. More than two-thirds of the six-day-olds tested 'turned toward their mother's pad, as did more than three-quarters of the eight- to ten-day-olds. Young babies prefer the familiar to the unfamiliar: here they recognised their mother's odour, and turned toward it. Although babies less than six days old did not turn toward their mother's pad, the older babies certainly did smell a difference – a difference that Macfarlane, when he smelled the pads himself, was unable to detect … Macfarlane's babies did not just detect an odour, they recognised one. To recognise something requires high-level processing within the brain – some kind of conscious processing beyond the reflex-like processing of the midbrain … Macfarlane's study does make it look as though the newborn's sensitivity to smells is close to being adult.'[20]

¶ When did I first taste?

The foetus can taste at 14 weeks. By that time, all of the tasting mechanisms are in place. Swallowing can be seen via ultrasound. By the end of the first trimester, the foetus controls the frequency of its own swallowing in response to sweet or bitter tastes.[21] The foetus regularly swallows amniotic fluid. Thus, according to Dr Gary Beauchamp, director of the Monell Chemical Senses Center in Philadelphia, a foetus swims in a 'smorgasbord' of flavours – the sweetness of glucose, the saltiness of sodium, and the bitterness of its own urine. According to Dr Tiffany Field, director of the Touch Research Institute at the University of Miami School of Medicine, 'amniotic fluid is kind of brackish-tasting, and we have videos of a foetus grimacing when it swallows the fluid'.[22]

For nearly 60 years we've known that humans have a definite preference for sweet-tasting things over bitter-tasting ones.[23]

The newborn is very good at distinguishing tastes. They will stick their tongues out when they taste sour things. Babies whose mothers frequently ate garlic during pregnancy will show a specific preference for garlic-flavoured foods. We think that's because they associate the garlic taste with mum – and of course they love mum!

¶ How soon was I able to cry and laugh?

Audible crying of a foetus has been recorded as early as 21 weeks.[24] Laughing occurs much later at about six months of age. Smiling is something else again. Research by Dr Susan Jones at the Department of Psychology at Indiana University has firmly established that we all have the equipment to smile from birth. The most recent work by Dr Jones involves observations of eighteen-month-old toddlers. She has also demonstrated that babies stop smiling very quickly if no-one pays attention to them.[25]

¶ How soon was I right- or left-handed?

Research on foetal thumb-sucking behaviour shows that handedness starts in the womb. When foetuses are observed by ultrasound imaging, some foetuses already indicate a preference for one hand over the other as early as 15 weeks gestational age. Dr Peter Hepper led the team of researchers from the Foetal Behaviour Investigations Unit at the Queen's University of Belfast, Northern Ireland.[26]

¶ Was I born with more than five senses?

This concerns the fascinating phenomenon of synesthesia – the merging of senses. It may be that we are born with senses which are undifferentiated and only become separated into our recognised five senses of sight, smell, hearing, taste, and touch sometime after birth.

We learn in school that each sense is separate and relates to a particular body organ. We hear sounds with our ears and see sights with our eyes – never vice versa. Indeed, we are led to believe that our senses are distinct, individual, and completely independent of each other. Thus, although a person may be totally blind, they may still have excellent hearing. We are also taught to assume that a human being cannot see sounds, hear sights, or touch tastes. The very concept of someone doing so seems completely foreign.

Synesthesia refers to the process wherein an individual experiences 'confusion' of senses – as if the five senses were somehow merged or fused into one. Sounds are seen, sights are heard, tastes are touched, and so on.

A growing body of evidence shows that synesthesia can occur in adults, although it is extremely rare. Notable so-called 'synesthetes' include the composers Olivier Messiaen, Aleksandr Scriabin, and Nikolai Rimsky-Korsakov. In fact, one alleged adult synesthete, the anonymous Russian vaudeville 'memory-expert' known simply as 'S', was studied for almost 30 years by a Russian psychologist.[27]

Nevertheless, stronger evidence exists that we are all synesthetes as newborns. Our five senses are at first blurred and become more specialised as we mature and our sensory channels to and from the brain become fully developed.

Dr Robert Hoffmann from the Department of Psychology at Carleton University in Ottawa attempted to measure the speed at which sensory impulses reached the brain in newborn infants aged between one and three months.[28] In an experiment, he first attached electrodes to the infant's head, exposed them to flashing lights from translucent panels, and recorded the speed of their electroencephalogram (EEG) waves. Readings were taken from directly over the visual cortex as well as from three areas of the brain far removed from the visual cortex. What happened then was as follows:

> These waves came through all four EEG electrodes: they welled up from all over the cortex. They were formed by energy from the eyes that was channelled throughout (and amplified by) the brain, where it would impinge upon, and mix with energy flowing through, other neuronal channels. Such impingement is the stuff of thought – not just verbal thought, but all associations

both meaningful and confused ... This showed that all of the babies did indeed perceive the stimuli directly – but these differences came not from the direct sensations; these differences came from the impingement of varying amounts of visual energy upon nonvisual areas of the brain. There energy from the eyes might mix with energy from the ears to produce vague sounds; or it might mix with energy innervating muscles to cause a twitch – which would in turn fire sensory neurons within those muscles, causing the sensation of movement. The amount of mixing depends on the amount of energy entering the nervous system. This total amount of energy added to the nervous system is a prime determinant of a newborn baby's perceptions.'[29]

Drs David Lewkowicz and Gerald Turkewitz from the Albert Einstein College of Medicine in New York discovered that babies between three and four weeks of age 'equate brighter lights with louder sounds'.[30] In an experiment, the two researchers first asked adults to attempt to adjust the volume of a loudspeaker to make it equal the brilliance of a light. The adults agreed remarkably well on the level of noise that seemed to equal the intensity of the light. Next, 20 infants were repeatedly exposed to the light while their pulse was monitored. A burst of noise was then substituted for one of the flashes. Although the level of noise that the adults had decided was equivalent to the light caused little reaction, every other level caused a marked quickening of each infant's pulse. And the quickening was proportionate to the difference between the intensity of the light and that of the matching sound.[31]

The two experiments suggest that the senses are intertwined in infants – that synesthesia exists in the infant's brain.

Can we learn to recapture in adulthood our previous state of synesthesia in infancy? And could we ever be taught to see without eyes or hear without ears by drawing upon our other senses? If so, then it may be possible to retrain people who have lost a sense to draw upon another.

Dr Richard Cytowic, a neurologist in Washington, DC, thinks that this just might be possible. Dr Cytowic is the author of the two best-known books on adult synesthesia.[32] He has seen over 40 synesthetes in his practice and is among the foremost authorities on the subject. According to Dr Cytowic, the sensory world of his patients is one of salty visions, purple odours, square tastes, and green wavy symphonies. One of Dr Cytowic's patients even allegedly has technicolour orgasms. Nevertheless, Dr Cytowic believes that understanding synesthesia will provide the key towards understanding the human mind.

What is the most common form of synesthesia?

That seems to be 'colour-hearing'. According to Dr Simon Baron-Cohen of the Department of Psychology at the University of London, such colour-hearing synesthetes nearly always have 'coloured vowels' and 'coloured letters'.[33] When they hear vowels or read letters, they report 'seeing' colours.

How common is synesthesia in adults? Of course it is rare. Approximately one in 20,000 people are synesthetes. Interestingly, 'when he [Dr Baron-Cohen] reported the [1987] findings in an interview on BBC Radio 4, over 200 women (and two men) wrote in, claiming to have synesthesia – an astonishing response, given that it was a science programme with at least equal numbers of male and female listeners'.[34]

Related to synesthesia, but very different, is 'blind-sight'. Some blind people possess a form of 'unconscious vision' that allows them to 'see' after a fashion – without being aware of it. This is 'blindsight'.

The term 'blindsight' was coined in 1980 by Dr L. Weiskrantz of the Department of Psychology at Oxford University and colleagues at the National Hospital in London. What the researchers discovered was this: 'Working with patients who had acquired visual defects as a result of damage to the brain cortex, they found that their subjects, if asked what they saw, said they saw nothing. In testing, though, asked to guess, they "guessed" right nearly 90 per cent of the time. This is far beyond the success rate accountable to pure chance. The researchers theorise that the kind of "sight" detected by these tests depends on a different neural pathway from the eye to the brain, one that passes through the midbrain rather than the cortex. In evolutionary terms, the midbrain is much older than the cortex. In a sense, it may be the animal part of our brain – an evolutionary holdover from our pre-human past.'[35]

Other experiments with blind people with 'blindsight' have found that some can catch a ball tossed to them – 'while insisting that they cannot see anything'.[36] In any case, deprived of vision, people with 'blindsight' somehow do seem to draw upon other senses to 'see'. Scientists have so far been unsuccessful in explaining such 'blindsight' abilities.

¶ How early can I conceive a baby?

Girls can begin conceiving at puberty. The average age of puberty has decreased by about two and a half months with each generation since the end of last century.

A Brazilian girl reputedly gave birth at six years, seven months, and three days of age. The oldest mother, from the US state of Oregon, gave birth at fifty-seven years, six months, fifteen days of age without the use of fertility drugs. However, modern drugs and artificial fertilisation techniques have pushed this date further and further up. Theoretically, there is no upper age limit on when we can no longer conceive children. But who wants to chase after a running toddler when you're 70?

¶ How many babies can I have?

Before the era of fertility drugs, an eighteenth-century Russian woman gave birth to 69 children of whom 67 survived into adulthood. She was able to do this by giving birth to 16 pairs of twins, seven sets of triplets, and four sets of quadruplets along the way.

Every human female possesses some two million eggs (ova) at birth. Of these, about 300,000 survive to puberty. And of these, only 450 are ultimately released for possible fertilisation – one each month during the reproductive years (roughly from 12 to 50 years of age). The human male produces half a billion sperm each day. Four hundred million are released in a single ejaculation. Men can remain fertile somewhat longer than women.

Assuming that a monogamous couple have sexual intercourse often enough so that all sperm produced are released, and assuming that a man remains fertile for fifty years, the chance of any one sperm fertilising any one ovum is 18,263,000,000,000,000,000 to one.

¶ Is 'immaculate conception' possible?

Reproduction without sperm is called parthenogenesis.
This can occur in some plants and in invertebrates.
It can also occur in some species of insects, fish, reptiles,
amphibians, and birds. Honey bees, wasps, and some
lizards are examples. But it does not occur in mammals –
including humans.

Experiments at Yale University have attempted to
induce parthenogenetic development in mice. In these
experiments, ova start to develop after exposure to these
three factors: electric shock, mechanical agitation, and
a saline solution. However, the embryo always dies prior
to the halfway point of gestation.

¶ How much did I weigh at birth?

Ask your parents about this one. Nine out of ten babies are born weighing between 2,400 and 4,800 g. Male babies are slightly heavier than girls on average (the difference is about 20 g). The heaviest baby on record weighed 13.15 kg at birth. The lightest baby to survive weighed a mere 283 g. For some unknown reason, babies born in November in the southern hemisphere and May in the northern hemisphere weigh on average approximately 170 g more than babies born in any other month.

Interestingly too, more babies are born between midnight and 8am than during the two other eight-hour periods. Tuesday is the most popular birthday, while Sunday is the least popular. More babies are born during days with a full moon than during any other time in the lunar cycle.

For unknown reasons, babies born to brown-haired mothers are delivered slightly faster than babies born to blonde-haired mothers.

¶ Are babies born without scars?

At birth, many newborns look like they've boxed 12 rounds with Mike Tyson. Babies who've been delivered with the help of forceps tend to be somewhat bruised. This heals very soon afterwards. However, in utero surgical procedures, which have now been undertaken around the world, have resulted in the discovery that the foetus does not scar.

¶ Why are babies born without teeth?

Ask any breastfeeding mother why nature builds this one in! Less pain to the nursing mother results in the baby being less likely to be rejected. Strangely, approximately one in every 2,000 babies has at least one erupted tooth at birth. Even more strangely, a large number of famous world leaders, including several emperors and dictators, are known to have already 'cut' a tooth at birth. Julius Caesar, Hannibal, Charlemagne, Napoleon, Mussolini and Hitler are among these. Could it be that the mother's pain during breastfeeding caused them to react negatively, withhold love, and even reject the child emotionally if not physically? In turn, did this rejection and denial cause the child to seek world power and domination later on as a substitute? It's fun to speculate. What would Sigmund Freud say?

¶ Will we ever be able to make an android to replace the human body?

We are near the time when we can replace human beings with robots in human form. The spectre of Arnold Schwarzenegger as 'The Terminator' or Lt Commander Data of *Star Trek: the Next Generation* is becoming less of a science fiction fantasy and more of a science fact.

According to one US scientist, 'the age of intelligent androids – not merely better computers, but a new form of life – may be as little as 20 years away'. Dr Maureen Caudill, a researcher at the Artificial Intelligence Laboratory at MIT in Cambridge, Massachusetts, has surveyed the achievements to date of various technologies required to make androids a reality. It is a new and fascinating research field. New advances occur almost monthly.[37]

Already, simple robots can navigate themselves around a room or pick up an egg without breaking it – a task once thought impossible even by experts. Far beyond this, there are now artificial neural networks closely approximating the human brain, along with sophisticated vision systems, memory systems, language systems, pattern recognition systems, and other processes. All of these processes would have to be replicated to a human level by any self-respecting android.

A robot may soon be carrying your baggage through the airports of the future.[38] Autonomous mobile robots equipped with sensing devices and artificial intelligence will soon perform a number of similar duties. Some potential robot jobs include house cleaners, office-mail carriers, minefield sweepers, and underwater or outer space–based construction workers.

Robots will soon be used in the travel industry to greet guests (in any language), babysit, and perform security tasks.[39] The night watchman may soon be a thing of the past. Admittedly, robots may not have 'the human touch', but they will work tirelessly and not complain when they are put in a closet during the off-season.

A new 'artificial muscle' is being made for robots. The muscle, made from a gelatin-like substance, has been developed by Dr David Brock at the Artificial Intelligence Laboratory at MIT. This invention may lead to smaller, stronger, and more flexible robots – bringing us another step closer to the age of androids.

More specifically, the Brock artificial muscle consists of polymer gel fibres. Polymer gels change their volume quickly in response to changes in pH. Thus, adding an acid solution causes the muscle to contract, while adding a basic solution makes it expand. By alternating doses of acid and basic solutions, the muscle can be made to lift or lower a 100 g weight.

This artificial muscle may someday replace many of

the gears, pulleys, and motors required in today's robots. In addition, robots could be free from dependence on electrical outlets because the muscle is operated solely by changes in pH.[40]

Dr Caudill describes a Japanese robot, the WABOT, that can read simple sheet music, play keyboard instruments, and accompany a human singer. And there is a machine-vision system, developed by German researchers, that has been able to successfully guide a car without human control at speeds up to nearly 90 kmph.

The key technology for building androids will be sophisticated neural networks. Dr Caudill explains that 'neural networks are information-processing systems physically structured in a way that mimics our current understanding of the brain'.

Neural networks differ from digital computer systems. Instead of following the instructions of a programme or pigeon-holing every bit of data in a specific location, as a digital computer does today, neural networks – like nerve cells that have many connections with each other – function by stimulation through patterns of activity. Dr Caudill argues that 'these patterns can be trained, rather than programmed, to achieve desired results'. Furthermore, 'neural networks also differ [from older technology] in other ways. Older artificial intelligence systems emphasise developing intelligent behaviour through logical step-by-step procedures, while neural networks are much less structured and much more flexible.'

Dr Caudill notes that neural network research is providing models to test how human brains work, learn, and organise themselves – the fundamentals of human psychology.

One experimental neural network, using a new psychological model, has successfully imitated the classical conditioning process by which animals are trained. This process was first described by Ivan Pavlov a century ago and represents a cornerstone of human psychology.

Another new neural network is modelling the internal 'mapping' that occurs when a newborn baby's brain absorbs its first environmental stimuli outside of its mother.

Thus, by trying to build an artificial brain – and ushering in the Android Age – we are also learning more about ourselves.[41]

¶ Can we ever live forever?

We humans may very well and very soon come very close to achieving immortality. Aging may become a thing of the past. And if aging is licked, then all the age-related diseases that kill us will also be licked. This would bring us very close to living forever.

As impossible as it sounds, genetically engineered drugs which will reverse the aging process will be available in the near future. Thus, it will be possible not only to halt the aging process but to turn the clock back. At age 70, we will be able to look and feel as we did at 50, and better than we did at 60. And should the age reversal process continue for another ten years, at 80 we could look and feel as we did at 40.

Age reversal is possible based on what scientists already know about the process of aging itself and about the already proven capacity of genetically engineered drugs to cure diseases. That same technology is being applied to aging.

Science tell us that we age because of problems with individual cells of our bodies. As it were, the whole is equal to the sum of its parts. Cells are continually being replaced by other cells throughout our lifetime. But the cells produced later in life are known to have more defects, mutations, and so-called DNA 'spelling errors' in them than the cells produced early in life. As time goes on, the proportion of these error-prone cells is larger than cells that are error-free. Our bodies do not look as good and we notice they do not function as well either.

Why do our cells commit more 'spelling errors'? This is probably due to the cumulative damage from free radicals within the cell. Free radicals are molecular fragments produced during normal cell metabolism that react wildly and unpredictably. Radiation, toxins, carcinogens, stress, and other factors can aid the production of free radicals. Furthermore, enzymes that serve as a kind of DNA 'spelling error' repair kit for cells are not produced in as great a quantity as before in the new cells our bodies produce when we are older.

Based on this idea, scientists are working to devise ways to boost and extend the life of the DNA-repair system. This would mean both the correcting of 'spelling errors' that can occur when DNA replicates itself when each new cell is born and the repair of DNA damage caused by free radicals. The reversal of aging would occur since error-prone, damaged, 'old' cells would eventually be replaced by error-free, undamaged, 'new' cells until the entire body is composed of only youthful cells.

Although it may seem a mere dream, this is no science-fiction fantasy. It is theory being turned into reality in laboratories around the world. At the Center for Molecular Science at the University of Texas Medical Branch in Galveston, Dr Samuel Wilson is undertaking experiments, which will lead to animal experiments and after that to human experiments, which will eventually result in drugs that humans can take to transform old age into youth.

Dr Wilson 'has already isolated the gene in mice for one of DNA's key repair enzymes in both humans and mice. The production of this enzyme is known to decline steeply with advancing age. His plan is to produce a line of mice that carry many extra copies of the gene for this critical enzyme, in the hope that those extra genes would keep turning out an abundance of the enzyme. These would be on hand to carry out their genetic repair work for much longer periods of time. Thus, the animal's DNA would accumulate errors and mutations at a much slower rate, and the mice would live to a venerable age.' According to Dr Wilson, 'I would guess it will be a year or more ... before we can report that we have a successfully engineered mouse. After that, it will only take another six months before we can tell for sure whether we have increased their life span.'[42]

Another team at Galveston is headed by Dr John Papaconstantinou. Research by Dr Papaconstantinou is still in its early stages. His group is working with several genes, including those involved in the response to every-day stress. Interestingly, the proteins these genes produce are able to manoeuvre themselves from the cytoplasm of the cell into the nucleus, where they can search out specific sites on the DNA and turn selected genes on and off. According to Dr Papaconstantinou, 'as we age some cells make too much of some proteins and not enough of others'. This means genes are being turned on or off inappropriately. Thus, 'our long-range hope is to learn enough about these processes to enable us to bring the cell back to its young-adult balance by manipulating the expression of the appropriate genes'.[43]

If Drs Wilson and Papaconstantinou, and others around the world are right in theory and successful in practice, their work could open the door to gene treatments promising age reversal – changing our cells – and hence ourselves.[44]

When will all of this happen? The best estimate is by the year 2010 – definitely worth living for.[45]

In any case, Dr Thomas Perls of the Harvard Medical School notes that 'traditional views of aging may need rethinking'. People are living longer now and are healthier than ever before. And it has been found that 'people in their late nineties or older are often healthier and more robust than those 20 years younger'.[46]

So there's every chance, more so than ever before and one way or another, that many of us may hit the century mark – and beyond.

2 · **The Brain**

We have learned much about the brain's many wonders, but there is still much, much more to learn about what Woody Allen once called his 'second favourite organ'.

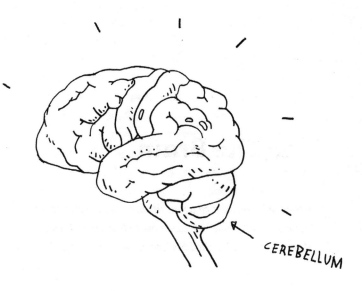

CEREBELLUM

¶ What is the brain anyway?

This is going to be a long explanation, but here goes.
The human body functions as an interdependent,
co-ordinated unit. Its activities, actions, and reactions
are directed by the brain through the vast and complex
network of the nervous system. The brain itself is the
most complex and largest mass of nervous tissue present
in the body. It directs the complex activities of the body.
Through the five senses – touch, sight, smell, taste,
and hearing – the brain keeps each of us in touch with
conditions and events in the world around us.

The human body performs two basic types of
movements or actions: voluntary and involuntary. In
a voluntary action, the brain calls upon the muscles or
organs of the body to perform a task. In an involuntary
action (a reflex), the senses communicate a condition or
situation to the brain (a stimulus), and the brain responds
by calling upon motor nerves to react or respond to the
stimulus. In a reflex, the brain is often by-passed. In such
a case, the sensory nerve may call upon the motor nerve
indirectly via the spinal cord for the required response or
directly via contact with the motor nerve. A reflex does
not require thought or 'brain work' as such in order for
the proper 'action–reaction' response to take place.

Most of the nerves that activate the muscles of the
body stem from the spinal cord. The spinal cord runs up

the spine and enters the skull cavity through an opening at the base of the skull (cranium). It then expands into the medulla and then to the cerebellum to form the base of the brain. Sitting upon the cerebellum are the brain's remaining major, distinct but interconnected parts which together fill the skull cavity: the pons, the midbrain, and especially the cerebrum.

The weight of the brain of the adult male is approximately 1.4 kg and the adult female about 1.25 kg. There is no significance in the gender difference in brain weight. Males generally have larger brains because they generally have larger bodies.

After conception, the brain of a human foetus appears to be electrically silent for about the first six weeks of life. After this time, intermittent 'slow wave' activity of low intensity occurs. In the human embryo, the brain first consists of three sections: the forebrain, the midbrain, and the hindbrain. As the embryo develops, the brain forms and develops its other parts.

Brain growth continues very rapidly up to the fifth year. It then continues more slowly until about the twentieth year. It remains stable in size throughout middle adulthood, while in advanced age it gradually loses weight.

The ability of humans to remember and utilise past experience, to cope with current situations, to think and reason, and to conceive of never-before-thought-of thoughts clearly differentiate humans from all other animals.

The cerebrum is the largest section of the brain. It is divided into two sections: the left cerebral hemisphere and the right cerebral hemisphere. The cerebrum is the centre of intelligence, sensation, emotions, and volition memory. It is divided into the left and right hemispheres by a long cavity called the longitudinal fissure. Each hemisphere is made up of an outer coating, the cerebral cortex (the so-called grey matter), covering an inner mass of white matter. But within each hemisphere is a space or ventricle connecting with the all-important third ventricle through an opening called the Munro foramen. These spaces within the hemispheres are collectively called the lateral ventricles. Moreover, each hemisphere has five lobes: the frontal lobe, the parietal lobe, the temporal lobe, the occipital lobe and the insula lobe. An injury to one cerebral hemisphere will affect the movement or action of the opposite side of the body.

The meninges are the three coverings which lie between the skull and the brain. Moving from the skull downwards:

The dura mater is the tough, fibrous membrane – rough on the outer surface but smooth on the inner surface – which protects the brain. It contains arteries, veins, and sensory nerves, and projects into the cranial cavity to divide it into partitions. The falx cerebri separates the cerebral hemispheres, the falx cerebelli partially separates the cerebellar hemispheres, and the tentorium cerebelli separates the cerebrum and the cerebellus.

The arachnoid membrane, the middle layer of covering, consists of bundles of blended fibrous and elastic tissue. Just below the arachnoid membrane is a space called the subarachnoid cavity which is filled with cerebrospinal fluid.

The pia mater is the innermost meninges protective layer. It consists of small arteries, veins, and connective tissue which help furnish the blood supply of the brain. The pia mater actually dips into the convolutions of the brain.

The lateral ventricles are spaces within the cerebral hemispheres of the forebrain which are also filled with cerebrospinal fluid. Long before birth, as the brain develops from the embryo to the foetus, the walls of the forebrain thicken and the cavity spaces shrink to form the thalami. Eventually, the cavity spaces of the lateral ventricles are reduced in size to a tiny slit known as the third ventricle.

The midbrain, as the central nervous system develops, has a thickening of its walls, which develop into two cylindrical bodies, the cerebral peduncles, and its central cavity also is reduced to a narrow canal.

The hindbrain includes the pons in the frontal section

above the medulla, or bulb in front of the cerebellum. The pons is made up of a connecting bridge of fibres which connect the halves of the cerebellum, joining the midbrain with the medulla below it. The medulla lies between the pons and the spinal cord and contains such vital centres as the respiratory, vasomotor, and cardiac centres. In the hindbrain is a large cavity, the fourth ventricle, which connects with the cerebral aqueduct above, and with the central canal of the spinal cord below.

The spinal cord contains nerve cells along its entire length and has bundles of long nerve processes, or nerve trunks, extending to different parts of the medulla, pons, midbrain, and cerebrum.

The cerebellum, the largest part of the hindbrain, lies in the posterior cranial fossa. It is covered by the layer of dura mater called tentorium which also serves to separate it from the posterior section of the cerebrum. It is composed of two hemispheres, with a middle section or lobe between them called the vermis. Bands of fibres, peduncles, connect the brain stem to the cerebellum. The superior cerebellum peduncle connects the midbrain and the middle cerebellear peduncles connects the medulla. The cerebellum functions as a reflex centre for co-ordination and degree of voluntary movements. Damage to the cerebellum affects co-ordination but as re-education of the motor areas occurs, it is clear that voluntary movements are co-ordinated in the cerebellum but do not originate there.

The corticle areas control particular motor, sensory and association responses.

The post central area is concerned with sensory activity, touch, and muscle. In the occipital lobe we find the visual centre, in the superior temporal convolution the auditory centre, in the hippocampal area the taste and smell centres.

The frontal area concerns association of ideas, conduct, behaviour, and intellectual concentration.

The precentral areas concern voluntary movement and exercise volitional control over the skeletal muscles. In the front of this area is a psychomotor area which is concerned with carrying out skilled acts.

The basal ganglia (or cerebral nuclei) are grey masses deep within the white matter of the cerebral hemispheres. The most important are the thalamus and the corpus striatum. The thalamus is oval shaped, in two parts which are separated by the third ventricle area, and connected with a stretch of gray matter termed the massa intermedia. It appears that the thalamus functions as a centre for crude and uncritical sensations and response, and in most animals it is the highest sensory response area, making their sensations primitive and imperfect.

In humans, the thalamus passes fresh relays of nerve fibre to the cerebral cortex, where finer interpretations and reactions enter conscious sensation. The function of the corpus striatum has not yet been clearly defined, but it seems to exert an affect on and steadies voluntary movement without initiating such movement.

The hypothalamus is found below the thalamus and forms the bottom and a portion of one wall of the third ventricle. It holds the temperature control centres of the body: one centre for controlling heat loss function through sweating and panting is found in the anterior (frontal) section, and the other section for preventing heat loss and increasing heat production is found in the posterior (rear) section. It also plays a part in the metabolic process through a connection with the posterior lobe of the pituitary gland.

¶ Can we live normally with only half a brain?

Some people have to live with only half a brain for their own health. For example, in serious cases of Sturge-Weber Syndrome there are major brain and body problems. There are port wine stain birthmarks on the face, especially around the eyes and forehead, pressure on the eyes resulting in serious glaucoma which can eventually leave a patient blind, epileptic-type seizures which can be very frequent, a lack of body co-ordination on one side of the body, learning disabilities, and mental retardation. According to Dr Steve Roach, a neurologist in Denver and consultant to the Sturge-Weber Foundation of Aurora, Colorado, when medications are ineffective in arresting the seizures, 'a [surgical] procedure to remove the hemisphere' causing the seizures has been available for several years. This procedure is called a hemisphere ectomy. Half the brain is surgically removed. Dr Roach adds that there is 'surprisingly little neurological impairment after the procedure'.

Dr Roach adds that a less radical procedure, but one less likely to work in stopping the seizures is called a corpus callosotomy. The hemispheres of the brain are surgically divided, but neither is removed.[1]

¶ Why are most people right-handed?

Odd that a chapter on the brain would discuss a matter of the hands, but such is the unusual character of this book. You see, your brain determines your handedness. The left side of the body is controlled by the right hemisphere of the brain and the right side of the body is controlled by the left hemisphere of the brain. Right-handers are left-hemisphere dominant while left-handers are right-hemisphere dominant. About 88 per cent of humans are right-handed. That leaves about 11 per cent left-handed. However, it depends on how handedness is defined. Some people favour one hand for doing some things and the other hand for other things. A truly ambidextrous person – one who is equally hemisphere dominant and who equally uses either hand – is quite rare.

According to science writer Marc McCutcheon, 'Most cases of left-handedness however, are thought to be caused by minor brain damage before or during birth. Many scientists believe the damage is due to reduced oxygen supply before birth.'[2]

Twins have a higher than usual rate of left-handedness. This is believed to be caused by twins having to share less space in utero and perhaps receiving less oxygen.

Sixty-five per cent of those suffering from autism are left-handed. Left-handedness is more common than usual among the world's artists and the world's gays.[3]

¶ Are identical twins either both right- or both left-handed?

Although identical twins are genetically the same, all sorts of things about their behaviour differ. Handedness is one of the ways behavioural differences can show.

'About 10 per cent of identical twins have different dominant hands', according to Dr David Lykken, professor of psychiatry at the University of Minnesota in Minneapolis.[4] Dr Lykken has conducted extensive research on the differences and similarities of twins.

He claims that 'nobody knows the causes of handedness. Right-handedness tends to run in families as does left-handedness. But this is not the whole story because many identical twins have different handedness.' He adds that 'other factors are presumed to be involved such as perhaps the position in the womb, but there are no definite answers'.

Fraternal twins develop from two fertilised eggs and are no more genetically alike or different than any other siblings. Identical twins develop from a single fertilised egg that splits to form two embryos and eventually two babies.

Dr Lykken maintains that when this split occurs 10 or more days after conception – after there has already been some cell division – then 'mirror twins' result. These are identical twins that are in some ways mirror images of each other. He explains that this 'is because the developing embryo has begun to develop laterality, with each side a little different'.[5]

¶ Why can't I tell my left from my right?

Welcome to the club – there are many members. If you often have trouble telling left from right, you may have what is termed 'handedness confusion'. Many adults experience difficulties and frustrations in distinguishing their left from their right in simple, daily activities.

Normally, children learn to discriminate between the left and right sides of their bodies at about age six or seven. Older children then expand this ability to include the left and right sides of other objects. As adults, this ability is essential to performing many ordinary activities such as driving and reading.

If an adult does not master this skill – and this seems to be all too often the case – then problems of confidence and competence emerge. Research shows that handedness confusion can last a lifetime. It is difficult to correct in adults, but less so in children.

Charles McMonnies, a Sydney optometrist and professor of optometry, claims that 'many children with learning difficulty, especially dyslexia, have persistent problems with right–left discrimination, including reversals of letters and words'.[6]

McMonnies says that 'it is important to teach children left–right body awareness early, not just to prevent problems in adulthood, but also to facilitate many aspects of early school experiences ...'.[7]

¶ Do animals ever show 'handedness'?

According to Dr Victor Denenberg, professor of bio-behavioural sciences and psychology at the University of Connecticut, many animal species exhibit handedness (or is it 'pawedness'?). He adds that, as in humans, handedness is clearly under brain control in animals. But unlike humans, a group of animals will normally split about 50/50 as to right- or left-handedness. Furthermore, some non-human primate research indicates that many use the left hand for simple tasks but the right for more complicated manipulations.[8]

Research by paleontologists Dr Loren Babcock of Ohio State University and Dr Richard Robison of the University of Kansas shows that ancient trilobites living 550 million years ago showed a handedness of sorts. Although these simple creatures had no hands, bite marks on fossils suggest an ancient propensity for turning to the right.[9]

¶ Who lives longer, right- or left-handers?

One of the most interesting debates among handedness researchers concerns whether or not handedness is linked to life expectancy. In 1988, Dr Diane Halpern of the California State University at San Bernadino and Dr Stanley Coren of the University of British Columbia presented findings that left-handers had shorter life spans.[10] But in 1989, Dr Max Anderson of the Canadian Statistical Analysis Service in Vancouver reported that left-handers had longer life spans.[11] Then, in 1992, Dr Charles Graham and colleagues at the Arkansas Children's Hospital in Little Rock concluded that left-handers do indeed live shorter lives. This is because, in an admittedly right-handed world, left-handed children have more accidents – some of which are fatal.[12] Equipment and mechanical devices are not designed with the left-handed minority in mind – often with disastrous consequences. Daniel Bristow from Kew, Surrey, writes in *New Scientist* that 'An interesting example is the SA-80 assault rifle. When fired from the left shoulder, it ejects spent cartridges, at great velocity, into the user's eye.'[13]

¶ Are we right- or left-headed just as we are right- or left-handed?

This question emerges from time to time especially in connection with controversies about the right and left hemispheres of the brain and their separate functions.

Medical and behavioural science evidence suggests that we are 'headed' just as we are 'handed'. That is, we are either right- or left-headed in much the same way as we are right- or left-handed.

Contrary to popular belief, the human skull is irregular in shape, oval, and thus asymmetrical. According to Dr Grange S. Coffin of the Department of Paediatrics at the University of California–San Francisco Medical Center, there are two common patterns of asymmetry of the human skull. Furthermore, one of these patterns appears much more frequently than the other.[14]

In the more common pattern, which Dr Coffin calls 'left-headed', there is a left-side and right-frontal prominence. It is as if a 'left-headed' person had placed their hands on the sides of their head and twisted the left half of the skull backward and the right half forward. In the less common pattern, which Dr Coffin calls 'right-headed' or 'reverse', there is a right-side and left-frontal prominence.

According to Dr Coffin's figures, 17 out of every 20 people are 'left-headed'. Moreover, he claims that 'headedness' is determined very early in life – before birth in fact. He believes that head shape is perhaps 'embryonically reflecting' the shape of the mother's uterus, maternal posture and sleeping position, the site of the embryo's implantation, or even 'obscure tidal and gravitational forces'.

¶ Why is 'water on the brain' so bad if the brain is normally surrounded by water?

'Water on the brain' is a very common birth defect and disease. Yet evidence suggests that there is much public ignorance about this condition and many misconceptions surround it.

'Water on the brain', or hydrocephalus, is not a rare condition at all. In fact, it occurs in about one in every 500 births. As such, it is more common than birth defects such as Down's Syndrome (about one per 700 births), spina bifida (about one per 1,000 births), or cystic fibrosis (about one per 2,000 births).

The so-called 'water' of 'water on the brain' is really not water at all. It is actually cerebrospinal fluid. This is the liquid which cushions and protects the brain and spinal cord from shock. This fluid is produced in the cavities of the brain called the ventricles. Normally, the cerebrospinal fluid continuously flows through the ventricles, bathes the surfaces of the brain and spinal cord, and is absorbed into the bloodstream.

However, in hydrocephalus this flow of fluid is seriously interrupted. The cerebrospinal fluid becomes trapped in the ventricles where it is unable to enter the bloodstream. The excess fluid causes the ventricles to expand, with the result that the brain becomes abnormally large. As a direct consequence of this larger brain, added pressure is placed upon the developing skull, and the baby's 'soft spot' (fontanelle) expands abnormally as well. If the pressure is not relieved early, permanent brain damage can result and the head may be horribly deformed.

More than 50 per cent of all cases of hydrocephalus are congenital – developing before birth. These cases probably are associated with infections that affect the foetus. Nevertheless, hydrocephalus can also begin during birth as the result of birth trauma or even during childhood as a complication of meningitis – an inflammation of the membranes that cover the brain and spinal cord (the meninges). Interestingly, in older children and even in adults 'water on the brain' can be triggered by a brain tumour or head injury.

Often, hydrocephalus occurs along with other birth defects. For example, nearly 70 per cent of children born with spina bifida also develop hydrocephalus.

One misconception about hydrocephalus is that it is always hereditary. Although there are occasions when the disease occurs twice in one family, the chance of this happening is no more than one in 20.

Another misconception is that there are proven ways to prevent hydrocephalus in the foetus. Although there are no 'proven' ways as such, there is some indication that the risk of birth defects such as hydrocephalus and spina bifida may be reduced through taking certain vitamins. For example, the 1991 newsletter of the Guardians of Hydrocephalus Research Foundation in New York carried a reprint of an article claiming that vitamins, particularly folic acid, may minimise risk.

Hydrocephalus can be detected through prenatal diagnostic techniques such as ultrasound. When it develops in the days, weeks, and months after birth, it can be detected through head measurements, skull x-rays, and CAT scans.

If diagnosis and treatment are early, normal mental and physical development is the most likely outcome. Unfortunately, if left undiagnosed and untreated, hydrocephalus can result in mental retardation, gross abnormalities, blindness, seizures, and muscular co-ordination problems.

Hydrocephalus is most commonly treated by surgically inserting an artificial tube called a 'shunt'. One end of the shunt is inserted into the ventricles and the other end is inserted into another part of the body, usually the abdomen. This allows the cerebrospinal fluid to drain off and be absorbed into the bloodstream.

'Shunt surgery', which in a child is normally con-ducted by a paediatric neurosurgeon, can relieve the worst symptoms of hydrocephalus and halt further brain damage or head growth. But it cannot reverse any brain damage which has already occurred. Further surgery may be needed as the shunt has to be lengthened periodically as the child grows. The child will probably always need a shunt. Shunt complications are few but occasionally there are infections and shunt malfunctions. If serious, the infections need to be treated and any malfunction must be corrected.

In the rare instances where 'water on the brain' occurs in an adult, the same diagnostic and treatment techniques are employed.

'Water on the brain' is not cured by a child's wet tears.[15]

¶ Can I get brain damage by simply drinking too much water?

The kidneys can only cope with so much water. Theoretically, if you continually drink water so that your kidneys can't deal with the flood, then your various body tissues can be swelled by the overflow. This would include brain tissue. It would be a condition called brain oedema. You could eventually die from it if you didn't stop. You would be literally drinking yourself to death.

¶ Can I easily put my thumb through a baby's 'soft spot' and touch the brain?

Let's knock this one on the head right away, so to speak. A baby's 'soft spot', located above their forehead, is called the fontanelle. Although a newborn's skull is not well formed at birth, the tissue covering the 'soft spot' is very, very tough.

¶ Can memories be injected from the brain of one person into that of another by using a hypodermic needle?

It is theoretically possible to transfer memories from one person to another – by injection.

In a number of successful laboratory experiments beginning some 40 years ago, so called 'memory molecules' have been successfully transferred via injection from one organism to another. Although the experiments so far have only involved simple life forms, the implications for human learning are immense.

Imagine acquiring the works of Shakespeare, Einstein, or last night's homework assignment, not through painstaking study but from a hypodermic syringe. And should memories fade with age, you could retrieve them from a vial as easily as a diabetic takes insulin.

The memory transfer experiments began in 1953. In that year, Robert Thompson and James McConnell, then graduate students in psychology at the University of Texas, attempted to 'teach' common planarian flatworms the simple task of avoiding an electric shock by swimming towards a light.

The planarian is a very simple creature and a favourite in laboratory experiments. It is about 3 cm long, found virtually everywhere in the world where there are ponds and streams, and the simplest animal to possess a true brain and a synaptic-type nervous system. Importantly too, it also reproduces both sexually and asexually. Sexually, one planarian may mate with another planarian and subsequently lay eggs to produce offspring. Asexually, a planarian may merely split in half. In the latter case, both the head and the tail re-grow to form two complete and genetically identical planarians. Each is at the same time its own mother, father, and twin sister and brother.

Just like Pavlov's dogs, the Thompson–McConnell experiments showed that planarians could 'learn' by simple stimulus-response conditioning. By 1955, Thompson and McConnell confirmed the then emerging theory of synaptic memory storage in the brain.

It had been theorised that memories are held somewhere within the synapses of the brain. A synapse is an extremely narrow, fluid-filled space between two nerve cells. Transmitter chemicals released by the end of one nerve cross the synapse and stimulate another nerve to fire. In humans, this process is not only the foundation of brain activity, but the basis of every human thought. Swedish biologist Holger Hyden had further theorised that, within the synapses, RNA (complex genetic molecules) actually held the memories. His experiments showed that the brains of trained rats were 'chemically different' from those of untrained rats – hence the notion of 'memory molecules'.

But could 'memory molecules' be transferred? McConnell, now at the University of Michigan, began the memory transference experiments in 1956. He, along with Daniel Kimble and Allan Jacobson, first trained planarians to avoid the shocks by swimming towards the light. Next, they cut them in half. After the heads and tails re-grew into complete planarians, they then re-tested both halves. They discovered that both the heads and the tails retained the memory of how to avoid the shocks. Thus, it seemed that not only do 'memory molecules' exist, but that they exist in multiples and can migrate.

In 1960, McConnell, Reeva Kimble and Barbara Humphries 'trained' some planarians to avoid shocks, chopped them up into tiny bits, and then fed them to another group of 'untrained' planarians. A second group of 'untrained' planarians received only normal food. Both 'untrained' groups were then tested with uncanny results. The cannibal planarians retained the memory of the training received by the 'educated' planarians they had eaten. Yet the non-cannibal planarians showed no evidence of any learning.

Subsequent McConnell team experiments involved merely injecting RNA of 'trained' planarians into 'untrained' ones to achieve the desired 'learning'.

From 1964 to the present, animal experiments in the US, Denmark, Czechoslovakia, and elsewhere reproduced these memory transference via injection results. Nevertheless, scientists still debate whether or not memories are encoded in RNA. Some dispute that memories were truly transferred from one animal to another in the experiments. Others point out that replicating such experiments is not always successful.

Recent experiments, such as those by Joseph Farley of Indiana University, pinpoint the true 'memory molecule' as protein kinase C (PKC). Protein kinase C seems to prime neurons to react strongly to a new stimulus. This line of work continues by a variety of researchers such as Terry Crow and James Forrester of the University of Texas at Houston with snails, and Aryeh Routtenberg of Northwestern University with chicks.

Of course, the animals and the learning tasks involved in these experiments are primitive. There is a great leap from a worm learning to avoid an electric shock to a human learning a Shakespeare play.

But what is fascinating about the memory transference experiments is that memories themselves emerge as real, physical things – able to move from being to being.[16]

¶ Can the human brain be fitted with an implant to give it the ability to directly access computer data banks?

The day of this possibility is definitely on the horizon. As technology advances, the human body will incorporate non-human elements. Humans may soon become 'composite beings' – part biological, part mechanical and part electronic. As one scientist puts it, humans will become 'metamen'.

Imagine surgically installing a tiny computer into your brain. That computer would have the capability to interface with your memory and thought processes and access entire libraries of information. You may one day have the ability to carry around in your head the contents of the British Library.

If you think this is science fiction or idle fancy, think again. Technological innovations are bringing this fancy closer to reality. Dr Gregory Stock, a biologist–physicist in Princeton, argues that the human 'composite beings' will not only be able to 'computer-to-brain' access entire libraries, but also be able to directly stimulate their brains in other ways. Mental problems will be a thing of the distant human past as humans can calm themselves, concentrate their attention, or feel pleasure with the help of computer implants.[17]

Furthermore, these future human brains will be so

powerful that their owners will view today's greatest geniuses as no more than simpletons.

Dr Stock maintains that the rest of the human body will also be transformed through its merging with machines. 'There will likely be not one, but many "human" forms in our future. As humans become more engineered, why would we not begin to manifest the same level of diversity seen in clothes, cars, and other designed objects? ... By the standards of the future, a multi-ethnic society ... will seem extremely homogeneous.'

When will all of this begin to happen? Dr Stock says that the beginning is already here. Computer technology is moving so fast that brain implants will be commonplace by the end of the next decade. 'Brain implants to boost thinking will be no more unusual than today's ear implants to boost hearing.'

According to Dr Stock, as metahumans become the rule rather than the exception, our entire civilisation will be revolutionised for the better in a way unprecedented in human history. He observes that barriers will tumble down as petty differences between people and nations will be rendered insignificant by the international community of advanced 'post-biological humans'. He writes, 'the middle of the ... millennium will reveal a Metaman, not battered and tenuously clinging to life as commonly portrayed in ... twentieth-century post-apocalyptic films, but rather healthy and growing, with human society thriving...'.

While Dr Stock shows how machines can be inserted into the human, Dr Hans Moravec shows how humans can be inserted into machines. Dr Moravec is the director of the Mobile Robot Laboratory at Carnegie–Mellon University in Pittsburgh. Dr Moravec has explained in detail how people could become robots, how they could download themselves into computers, and how all of this could be accomplished within the next 50 years.

Dr Moravec describes the character of this truly human robot à la computer. 'This thing could now carry on the life of the person whose mind you transferred to it. The robot would have all the same skills and all the same motivations as the human being did, and so it could raise children or do anything else the human could do. In fact, and for all practical purposes, this 'robot' is the human being ... Everything the human being did, this artificial replacement does too. So if you don't want to call it a human being, it seems like just perversity on your part.'[18]

Whatever the form, Dr Stock thinks that present biological humans will eventually learn to accept, even welcome the advent of post-biological 'metaman'. He notes that although social turmoil, uncertainty, and poverty will not disappear from the human landscape in the next few years, 'humanity's trajectory is toward a rich and vital future'.

He never 'metaman' he didn't like.[19]

3 · The Head

In A Shropshire Lad, *poet Alfred Edward Housman (1859–1936) wrote, 'Empty heads and tongues a-talking, Make the rough road easy walking'. No, Housman was not predicting the plots of TV soap operas. But as he mentions 'heads' – the subject of this chapter (we each have one, you know) – his lines provide a somewhat awkward introduction to what lies below.*

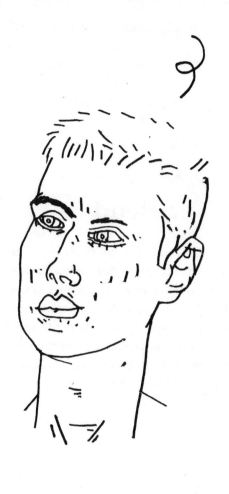

¶ Does a bigger skull mean a brighter person?

People have been asking this question for more than 200 years. Many myths surround it. An entire field of pseudo-science called phrenology was built upon the contention that skull size, shape and bumps determined intelligence, personality, and even one's position on the evolutionary tree. Phrenology reached its greatest level of popularity in the late nineteenth and early twentieth centuries.

We can explode forever the myth that the larger a person's head, the more developmentally advanced that person will be. This was confirmed by researchers, headed by Dr Teresa Brennan of the Department of Paediatrics at the University of Virginia Medical Center in Charlottesville, examining the relationship between 'relatively small heads' in children and later development.

The researchers used standardised developmental tests, including the Stanford-Binet IQ test (given at four years of age) and the Wechsler Intelligence Scales (given at seven years of age). The Brennan team found no developmental differences between the children with relatively small heads and children with average or relatively large heads.[1]

Related research found that narrow-headedness makes no difference in a child's eventual developmental outcome either. The research team that made this finding, studying preterm infants, was led by Dr Alison

Elliman of the Queen Charlotte's Hospital for Women in London.[2] Too bad that hat size does not correlate with intelligence. If it did, then intelligence testing could be done with a tape measure.

¶ Can you bore a hole in your head to relieve a headache?

This body mystery question arises for several reasons. First, when you have a headache, the pressure builds inside your head. People often fantasise that their head would feel better if the pressure was relieved. So why not drill a hole in the head? Second, it is a historical fact that the ancient Egyptians performed such operations to alleviate pressure on the brain due to stroke and other brain-based problems.

Let's look at this.

The most humorous report on 'head-boring' is found in the August 1986 issue of *The People's Medical Journal*. According to this article, 'head-boring', a fad long since thought to be dead and buried, has made somewhat of a re-emergence. The report was based on an article in the April 1986 issue of *MD*.[3]

In 1962, Dr Bart Hughes, an experimental neurologist in New York, proposed the unorthodox and highly suspect theory that one's state of consciousness was dependent on the volume of blood in the brain. He advanced the idea that the brain was usually 'constricted' by the skull. Therefore, the best way to 'raise conscious-ness' was to drill a hole in the skull. This is called self-trepanation.

Self-trepanation had a modest flock of true believers

in the 1960s and one disciple even wrote a book on various head-boring techniques. The author, W. Smith, entitled the book simply *Bore Hole*.[4] Perhaps it was meant to be a book on well digging? Another follower, a young woman by the name of Amanda Fielding, gained a tiny slice of medical immortality of sorts by filming her own trepanation with an electric drill. Needless to say, head-boring did not catch on in the 1960s.

Interestingly, the head-boring fad may be making a modest comeback. Beginning in 1984, some US doctors began receiving inquiries from patients about this form of 'therapy'. It also seems that some have been encouraged by the fact that head-boring was mentioned in the enormously popular Hollywood movie *Ghostbusters*.[5]

¶ Can you keep a severed head alive?

Medical science is now capable of keeping a severed human head alive.

In 1988, the US government granted a patent for a perfusing device that would keep a severed head alive after being surgically removed from the body. According to the holder of the patent, with his device and today's drugs which eliminate blood clots and other waste products from the brain, a severed human head could be kept alive indefinitely.

The procedure calls for the head to be surgically severed from the body at the top of the neck, mounted upright, and attached to the perfusing device. The device consists mostly of plastic tubes connecting the bottom of the head and neck to circulation machines which continue brain maintenance.

Perfusion refers to the process of artificially keeping brain-sustaining oxygen, blood, fluids, and other elements circulating properly. Using the perfusion device would allow the brain to think, the eyes to see, the ears to hear, the eyelids to shut during sleep, and other head and brain functions to take place.

Chet Fleming, an american molecular biologist, engineer, and patent attorney, holds the rights to the perfusion device – US patent 4666425. The patent was granted on the basis of blueprints only. This is called a 'prophetic patent' as there is no working model. Mr Fleming plans to build his device and make it available to experimenters.

Mr Fleming writes in the *British Medical Journal* that 'the technology for perfusing a severed head has important potential advantages, for research and for prolonging life in a conscious and communicative state with, probably, less pain than many dying people suffer today. The difficult question is whether the advantages outweigh the disadvantages and dangers.'[6]

Mr Fleming maintains that the head-severing operation and use of his invention will definitely have some takers. He writes, 'I have been contacted by half a dozen people who want to know how soon the operation will be available and how much it will cost. Some are dying; others are paralysed. Most said that if the mind remains clear and the head can still think, remember, see, read, hear, and talk, and if the operation leads to numbness rather than pain below the neck then they would want it.'

In 1988 Mr Fleming privately published a book entitled, *If We Can Keep a Severed Head Alive*.[7] In this treatise, he outlines the reasons for his efforts on behalf of the perfusing device. He claims that one of the reasons why he obtained a patent is to protect this technology from falling into the wrong hands. Moreover, he promises to make his device available to 'any scientist or surgeon who wants to try it' on either animals or humans. But if they do, they must first consult three independent review panels such as those already existing in major universities.

One panel would be 'animal care committees' which would control experiments on animals. Another would be 'institutional review boards' which must approve experiments on human subjects. The final panel would be 'institutional biosafety committees' which serve to control genetic engineering experiments.

Animal experiments wherein the head is severed have been undertaken since the beginning of this century. In a notable 1907 experiment, the entire upper half of a dog was transplanted. For this work, the French physiologist, Alexis Carrel received a Nobel Prize in 1912.

Yet perhaps the most famous experiment of this kind took place in the early 1970s. At a 1971 conference of surgeons meeting in New Haven, Connecticut, five doctors from the Case Western Reserve University in Cleveland, Ohio, presented evidence that they were able to keep rhesus monkey heads alive for up to 36 hours. The leader of the team, Dr Robert White, claimed that the monkeys remained fully conscious. He remarked, 'our animals were not sleepy: they tracked; they ate; and they bit you if you brought your delicate hands near the mouth'.[8]

According to Mr Fleming, the reason that the White

team's monkeys died was 'because of heparin overuse, a problem which can be overcome today using an extracorporeal heparin remover'. Heparin is a blood anti-coagulant. Mr Fleming adds that 'research is being done today on intact brains which continue to generate brain waves after the sensory organs and the skull have been cut away'.

The perfusion of severed human heads raises enormous ethical, legal, as well as medical questions.[9] And if there are indeed those who would choose this operation, it proves once again how desperately humans will grasp at any hope of life continuing – in any form.[10]

¶ Can you re-shape your skull?

We all know that we can change our mind, but less well known is that we can also change our skull – or at least others can do it for us.

Of course, the human skull in its normal, natural state is perfect for its purpose and needs no change. In fact, it is a marvellous piece of engineering. Contrary to popular belief, the skull is not one bone. Actually, there are 22 bones that make up the framework of the human head – not counting the teeth. The average skull in adulthood is about 21.59 cm high, 17.78 cm long and 15.24 cm wide. The height of the skull is often used as the basis for estimating the measurements of the rest of the body. The eight sections of the skull that encase the brain, called the cranium, protect it wonderfully. Besides the brain, the skull also protects the eyes while maximising our ability to see. And it protects the inner workings of the inner ear so that we could hear (although not as well) even if our outer ears were sliced off. Interestingly, the bones of the cranium have ragged edges which interlock with the edges of adjoining bones more finely than any jigsaw puzzle. These edges, called cranial sutures, follow no regular pattern and are as individual to each of us as our fingerprints.

There are of course many other interesting facts about the human skull, but the question remains, why would anyone want to change this wonderful artwork?

A Sydney doctor was recently asked by a new mother, 'Doctor, can you change the shape of my baby's head? It is so ugly!' The doctor reassured her that her baby was normal, that many babies are born with somewhat misshapen heads, and that as a child grows, the head takes on 'a more pleasing shape' in proportion to the rest of the body. In other words, there was nothing for her to worry about. Nevertheless, his answer was unsatisfactory to the mother – she wanted her baby's head to be beautiful now.

In fact, there was something the doctor could do, although it would be unethical for him to do it – and humans have being doing it for thousands of years. It is called head shaping, head moulding, or 'intentional cranial reformation'. In the field of plastic surgery, it is called 'non-operative cranioplasty'. In any case, it comes to the same thing – changing the shape of the skull to fit some desired image.

In fact, head shaping by various means is a practice that 'dates back to very ancient times and has enjoyed wide acceptance' in many societies. This is according to Dr F.O. Adebonojo, a plastic surgeon from East Tennessee State University, writing in the *Journal of the American Medical Association*.[11]

Head shaping is known to have occurred in Europe, Asia, Africa and in both North and South America.

Interestingly, it was not practised by ancient Aboriginal Australians.

Head shaping dates back to at least 2000 BC on both the Mediterranean islands of Cyprus and Crete as well as in ancient Egypt.

In order to successfully mould the skull, it must be done in infancy when the bones that make up the human skull are still soft and pliable. In most cultures where head shaping was commonly practised, the moulding process began within a few days after birth. External force had to be applied in some way. For example, among the Kwakiutl Indians of British Columbia, long heads were viewed as beautiful. Thus, the infant's head was sandwiched between two slats of wood. The slats were then tightly bound together with twine. This elongated the head somewhat. After a time, the head was re-bound more tightly, thus producing further elongation. The process was continued for at least three months, but sometimes much longer, until the desired degree of elongation was produced.

Among the ancient Incas of Peru, a naturally long head was made even longer, while a naturally short one was made shorter. Inca notions of beauty at the time dictated that natural characteristics should be emphasised and exaggerated, whatever they were.

Sometimes people re-shaped the heads of their children for reasons other than beauty. For example, only the royal class was allowed to shape their heads in both Tahiti and Hawaii. Thus, members of the ruling class could be easily distinguished from everyone else.

Sometimes too, head shaping was employed for only one gender. For example, the famous ancient Greek physician Hippocrates (c. 460–377 BC) reports that female childhood 'head compression' was used among the Greek aristocracy. It has been speculated that this was done both to make women prettier for men and impair them intellectually – as it were, compression serving oppression. If so, then the head shaping of women among the ancient Greeks served much the same function as the foot binding of women among the ancient Chinese.

Among societies practising head shaping, Dr Adebonojo writes, 'the range of intentional skull deformation (reformation) is quite large, varying from one extreme that produced grotesque cranial distortions to mild or minor deformations'. Nevertheless, Dr Adebonojo adds, head shaping appears to be a fairly safe practice 'since it does not appear to change cranial vault weight or volume or intellectual capacity'.

Today, there is a definite place in modern medicine for various forms of head re-shaping. As Dr Adebonojo writes, 'some plastic surgeons have, in fact, considered the use of techniques similar to head shaping or moulding as a valuable tool in the correction of certain craniofacial abnormalities'. A classic article on this topic appeared in a 1973 issue of *The Lancet*.[12] Three New York University Medical School surgeons, led by Dr F. Epstein, describe the treatment of five young patients with hydrocephalus. They write that their treatment is 'based on the premise that increasing the resistance to expansion of the skull by compressive bandaging promotes increased CSF [cerebro-spinal fluid] absorption...'.

Furthermore, 'non-operative cranioplasty' is sometimes used to treat children with skull and facial birth defects of various kinds. For example, Hemifacial Microsomia/Goldenhar Syndrome (HMGS) is the second most common birth defect occurring in the Western world today. Approximately one per 3,500 births are affected. Usually the eye, ear, cheek, jaw and one side of the face are deformed. Although HMGS is extremely variable from child to child, some children have cleft palate/lip, organ and system problems, and a minority suffer mental retardation. Such children often require surgical reformation of the skull. But the techniques today are somewhat more sophisticated than two slats of wood and twine.[13]

In any case, just as we humans have been known to change our mind, we also have been known to change our skull.[14]

¶ How do headhunters shrink a skull?

Headhunting is a curious custom found in many parts of the world. However, headshrinking is found only in some regions of South America.[15] Headshrinking is a very elaborate process which is filled with cultural, symbolic and religious significance.[16]

According to Dr Jim Leavesley, a retired general practitioner in Margaret River, Western Australia, although techniques vary somewhat, 'the result after hours or even weeks of dedication and ritual was a head about the size of a large orange'.[17]

After the enemy is killed, the skin of the neck is cut low onto the chest, the neck is separated with a sharp blade and the head is removed. A strip of bark is then passed through the mouth, down the oesophagus, and out through the end. This is used to hang or carry the head.

A pot is then filled with water and brought to the boil. The skin on the back of the head is slit up to the crown of the head and teased away from the bone in order to be peeled completely forward. Eventually, the skull and lower jaw can be removed totally and are discarded.

What remains of the eyelids are then sewn together from the inside. Small pieces of wood are then put through the lips and the mouth is bound shut.[18]

A remaining mop of skin, scalp, and hair is dumped

into the boiling water with special leaves for two hours. Juices from the leaves supposedly keep the hair from falling out.

The skin over the back of the head is then brought together so as to resemble a deflated human head.[19]

Rounded stones are then heated in the fire. When hot they are levered into the severed neck opening.

The head is then rotated by hand while another hot stone, held in a leaf, is used to smooth out the external features to resemble a human appearance.

The head begins to contract as it cools. Smaller stones and hot sand are dribbled into the head via the neck to permit all parts to shrink evenly. The hairs of the face, eyebrows and eyelashes are now disproportionately long, so they are singed off. The head is then hung about a metre above a fire where it smokes overnight.

Later, its natural oils are buffed to give it a permanent lustre. The pieces of wood in the mouth are replaced with coloured threads which are cut to the length of the hair.

It is believed that if the mouth were to remain open, the dead enemy could curse its slayer.[20]

¶ Why do the heads of baby chimpanzees closely resemble the heads of baby humans but the heads of adult chimpanzees differ greatly from the heads of adult humans?

If you were to shave the face and hair off a baby chimp and keep the rest of its body wrapped up, it might pass for a human baby if someone wasn't looking too carefully. Then again maybe not, since a chimp's ears are pretty big. In any case, baby chimps and humans do look remarkably alike as animals go. It is also true that adult chimps and humans do not look at all alike. Why the change?

Baby humans possess a round cranium with both a flat nose and jaw at birth. So do baby chimps. In fact, human and chimp embryos and foetuses look even more alike than they do as newborns! The human newborn's brain grows rapidly, whereas the rate of brain growth for the newborn chimp starts to slow down quite a bit. As a chimp grows, the jaw juts out, the nose stays flat, the teeth grow larger, and the superciliary ridge (the bone beneath the eyebrows) becomes more prominent. The cranial vault is lower and smaller than in humans.

According to Dr Stephen J. Gould, famed paleo-biologist at Harvard University, the difference between humans and chimps is that the human brain grows at a faster rate for longer. The human skull must accommodate this bigger brain – like the glove must fit the hand.[21]

4 · The Eyes

So much has been written about eyes. Poets call the eyes the windows of the soul. 'Beauty is in the eye of the beholder' is a saying attributed to Margaret Wolfe Hungerford (1855–1897). 'Night hath a thousand eyes' comes from John Lyly (1554–1606). 'You've Got Reptilian Eyes (But I Still Love You)' is a song from Woody Allen's movie, Zelig (1983). The immortal lines are endless.

So much has been written about eyes, yet most of us know very little about them.

¶ Can some people pop their eyes out of their head?

This is a very weird ability – and very rare. The first case in the medical literature of a person having the ability to 'pop' their eyes in and out without any apparent injury or discomfort was reported in the *American Journal of Ophthalmology* way back in 1928.[1] Dr H. Ferrer described a twenty-year-old male 'who had the ability to dislocate at will either eye separately or both simultaneously'. Four years later, Dr J. Smith reported in the *Journal of the American Medical Association* that an eleven-year-old boy could do the same thing.[2] Every so often an article appears about someone able to perform this feat.

The late actor–comedian, Marty Feldman, had eyes that looked like they popped out of his head, but he could not pop them in or out. His face just looked that way all the time. Actually, Feldman suffered from Crouzon's disease which is sometimes called craniofacial dysostosis. In Crouzon's sufferers, the eyes look as if they are popping out of their sockets although vision is not impaired. Feldman was no exception. On his passport, where it asked for 'any distinguishing physical characteristics' Feldman simply wrote 'face'.

Dr Barnett Berman of Baltimore calls the ability to voluntarily propel the eyes 'the double whammy syndrome'. He writes about this weird ability:

'In conclusion, I quote Emerson, who wrote, "Some eyes have no more expression than blueberries, while others are as deep as a well, which you can fall into." Once observed, the double whammy leaves an indelible impression that is unforgettable.'[3]

¶ Why do I cry?

Research suggests that we cry for both physiological and emotional reasons. In fact, crying may be important to maintain both physical and psychological health.

Everyone knows that crying is an emotional release and relieves pent-up stress. But what is not as well known is that the tears from crying are almost certainly one way for the body to cleanse itself of toxic substances. For example, salts are excreted in tears just as they are through sweat and urine. Tears contain a variety of different salts which come from the diet via the blood. Salt in food is absorbed by the intestines and enters the blood stream. As blood flows through the tear-producing lachrymal glands, the salt enters tears.

It has been known since the French chemist Antoine Lavoisier (1743–1794) conducted the first scientific study of tears in 1791 that tears contain sodium chloride (salt). They also contain other salts such as potassium chloride, plus factors which assist salt formation. Among these are calcium, bicarbonate, and manganese. Experiments held more than 30 years ago showed that the concentration of sodium in tears was the same as that in blood.

There may be a strong element of truth in the expression 'a good cry makes you feel better'. The ancient Greek philosopher Aristotle (384–322 BC) theorised that crying at a dramatic performance benefits a person by a process called 'catharsis' – the reduction of stress through emotional release. This word has figured its way prominently into our modern vocabulary of psychology. In a classic 1906 article in the *American Journal of Psychology*, Dr Alvin Borgquist found that 54 out of 57 patients reported positive health benefits after crying.[4] Recent studies consistently report similar findings.

At the Ramsey Dry Eye and Tear Research Center in St Paul, Minnesota, biochemist Dr William Frey has found that 'emotional tears' produced by 'tear-jerker' movies differ in chemical content from 'irritant tears' produced by breathing onion juice fumes. Dr Frey has found that 'emotional tears' contain more protein compared to 'irritant tears'. However, the significance of this finding remains unclear.[5]

In a study at the University of Pittsburgh School of Nursing, psychiatric researcher Margaret Crepeau found that, among 137 men and women, 'healthy people are more likely to cry and have a positive attitude toward tears than are those with ulcers and colitis – two conditions thought to be related to stress'.[6]

Researchers are currently investigating the content of tears for substances such as endorphins, ACTH, prolactin and growth hormone, which are all released by stress.

Furthermore, the average cry lasts approximately six minutes. A typical one-year-old infant cries 65 times per month.[7]

¶ Does any other animal cry?

The human being is the only primate that cries. Only one other land animal cries: the elephant. Other marine animals that cry include seals, sea otters and saltwater crocodiles (the so-called 'crocodile tears'). All of these animals cry only to get rid of salt. However, one scientist, Dr G.W. Steller, a zoologist at Harvard University who has studied sea otters extensively, thinks that sea otters are capable of crying emotional tears. According to Dr Steller, 'I have sometimes deprived females of their young on purpose, sparing the lives of their mothers, and they would weep over their affliction just like human beings.'

¶ Do women really cry more often than men?

It is often asserted that men rarely, if ever, cry – especially in public. The conventional stereotype is that crying indicates weakness in a man. The strength of this stereotype in forming opinion is exhibited from time to time.[8] In 1968, US Senator Edmund Muskie, then the leading candidate for the Democratic Party nomination for president, saw his hopes dashed when he was filmed 'sobbing' while addressing a crowd. Senator Muskie later denied that he had cried. Instead, he attributed his tears to the near-freezing weather conditions. He may have been right, as exposure to intense cold can cause tearing, but the public refused to believe him. He was dismissed as 'too weak to be president' and eventually dropped out of the presidential race and into obscurity – hardly anyone, pardon the phrase, shedding a tear.

In recent years, there is some evidence that Australians are a little more likely to allow their male leaders to cry. The former Australian Prime Minister Bob Hawke is a case in point. He has cried in public on several occasions, yet remains the second longest-serving prime minister in Australian history.

Nevertheless, evidence suggests that men cry more often than most of us are ready to admit. For example, US surveys show that women average 5.3 cries per month compared to 1.4 cries for men. That is nearly 17 cries per year for the average man.[9]

If crying is a vital factor in health maintenance by releasing emotions and stress and if women cry more readily than men in our society, this may help to explain why men are more ravaged by stress-related diseases and die earlier.

Perhaps women know that a tear a day keeps the doctor away?[10]

¶ Why do tears taste salty?

As previously mentioned (see page 106), tears contain salt. Tears are in fact approximately 0.9 per cent salt. It would be impossible to mask this taste. Anecdotal evidence from behavioural science indicates that most people cannot distinguish the taste of tears from that of unpolluted sea water. Of course, finding unpolluted sea water nowadays is another matter.

¶ Where do tears come from?

Above and slightly behind each eye, under the frontal bones of the skull, we have an almond-shaped lacrimal gland. About a dozen or so channels (lacrimal ducts) run from the lacrimal gland to the eye and eyelid. When we blink, the lacrimal gland is stimulated and tears wash over the eye. The eye is kept perfectly moist this way. It is also kept very clean. Tears are sterile and contain bacteria-destroying enzymes which protect against infection.

¶ Where do tears go if not down my face?

When we cry, some moisture is lost through evaporation but most of the moisture from tears drains from the inside corner of the eye, down through two other channels (lacrimal canals) into the peanut-shaped lacrimal sac, and finally into the nasolacrimal duct where it drains into the nasal cavity. This is why when you cry a great deal, your nose runs too.

¶ Why do I sometimes cry at a happy ending?

Most psychologists maintain that it is a myth that we cry when we are happy. Actually, there is no such thing as 'tears of happiness'. It is not out of happiness that we cry, but because unpleasant feelings are stirred up. Both men and women can suppress the urge to cry. During the emotional experience of a 'four star, four hanky' movie, we may hold back the waters until the film's climax. We then have a climax of our own, so to speak – tears gushing forth. The energy used to hold back the flow is now discharged. These tears are an expression of this emotional release with combined feelings of anxiety, fear, sadness, relief, and so on.

It is not only in movies that we seem to cry at happy endings. We emotionally release with tears when someone close to us is greeted after a long absence, when someone close emerges alive and well after delicate surgery, when someone close survives an accident, and at other times. Such emotional occasions are endless. And we never run out of tears.

¶ Why do I blink?

Studies show that humans blink every two to 10 seconds on average. We do this automatically as our eyelids protect our eyes from injury. Occasionally, we blink voluntarily as part of our body language. For example, we wink or 'bat an eyelash' in order to send flirtatious messages to another party. We also blink to express astonishment as if to say, 'I can't believe my eyes!' Non-human primates such as baboons have a white region of the upper eyelid. When they blink it is a threat gesture and signals both an alarm and a warning.

In humans, the inner surface of both the upper and lower eyelids, as well as the eyes themselves, are covered by a thin and almost transparent membrane called the conjunctiva. The conjunctiva functions to keep the surface of the eye moist, as moist eyes are necessary for proper eye movement. During blinking, the eyelid spreads moist tears over the conjunctiva and also into the corners of the eye.

¶ Why can't some people control their blinking?

Sometimes a person cannot control eye blinking. If so, they may have a medical condition called blepharospasms – uncontrolled eyelid movements.

No-one knows what causes blepharospasms, how to cure them, or how many people are afflicted by them. Few doctors even knew what it was a decade ago but research is coming up with some interesting clues to understanding this condition.

Dr John Hotson, chief of neurology at the Santa Clara Valley Medical Center in San Jose, has long been interested in the way blepharospasm patients see and remember action. Dr Hotson believes that the way the eyes see movement can serve as a clinician's window to pinpoint the defect at the root of blepharospasms. If he is successful, his work would mark a step toward developing a diagnostic technique and cure for this bizarre ailment that turns patients' lives upside down.

Since eyes normally blink once every few seconds, the lids close so gently and quickly that vision is not interrupted at all. In many people, an irritation or a tic may cause the lids to flutter or shut unusually hard. However, 'blefros' (as they are sometimes called) suffer spasms of a different kind. Their eyebrows yank down. Their lids jerk and slam shut. Over time, the spasms spread. The jaw may lock, the face may twitch, and the neck may be gripped in pain.

Although such patients are often rudely dismissed as having a psychiatric-based problem, blepharospasm experts such as Dr Hotson are convinced that the problem is physical and not psychological. This seems to be confirmed by his research which suggests that patients reflect a typical range of personality types and perform normally on psychological tests. Moreover, according to Dr Hotson, the ailment progresses in predictable patterns similar to those which doctors often see in patients with Parkinson's disease.

Currently, two operations can ease the symptoms of blepharospasms. In one, the surgeon makes an incision near the ear, pulls down the forehead and severs the branches of the nerve that controls eyelid movement. The procedure is delicate and often disfiguring. The second, a more common surgical option, involves the surgical extraction of the squeezing muscles in and around the upper eyelids.

Interestingly, many blepharospasm patients rely on experimental injections of botulinum toxins – the same poisons that cause botulism. Nearly 300 doctors world-wide are testing this therapy. Patients receive a series of injections in their lids and brows to paralyse the muscles. The injections appear to be effective but wear off after only a few months and thus must be repeated.

Dr Hotson hopes to find a longer lasting remedy.

As the first step in this, he is searching for the cause of blepharospasms. He theorises that wild blinking reflects miscommunication among nerve cells. He suspects that the errant cells are located in the brain's basal ganglia (a major centre for controlling movement). Basal ganglia pathways affecting eye movement is the factor currently receiving Dr Hotson's closest attention.

According to Dr Hotson, research shows that a pathway abnormality can subtly hamper a person's memory for movement. For example, a person could watch a dot shoot across a computer screen and think he or she recalls the direction of movement perfectly. However, they would probably be unable to trace the line of movement as accurately as could someone without this brain disorder.

Dr Hotson is looking for evidence of such an impairment in blepharospasm patients. He is having research subjects watch dots dance on dark screens. Their speed and accuracy in 'charting' is carefully being monitored. The research will continue for a few more years at least before firm results can be seen.

After that, we shall see.[11]

¶ Am I happier the more often I blink?

It is not often expressed by medical or behavioural scientists but it may be that the happier we are, the more often we are likely to blink. This is the firm assertion of one noted US neuroscientist. Furthermore, he claims that research shows that eye blinks offer clear signs of emotional wellbeing.

According to Dr Joseph Teece, a professor of neuropsychology at Boston College, 'pleasant feelings mean fewer blinks, while negative ones such as anxiety and pain get those eyes afluttering'.[12]

To illustrate his point, Dr Teece claims that during the 1988 US presidential debates between George Bush and Michael Dukakis, 'George Bush's average rate of 67 blinks per minute went to 89 when talking about abortion and to 44 when praising Dan Quayle.' According to some body language experts, eye blink rate is a predictor of who will win an election where television image is critical. For example, in the 1992 US presidential debates, Bill Clinton had a somewhat lower eye blink rate than George Bush, but the difference was small – as was Clinton's margin of victory.

Eye blinks are already being used in health settings. For example, some dentists are using eye blink rates to track when patients are feeling pain. And psycho-therapists are using them to tell when patients are uncovering painful emotions.

It may be that blinking may have medical diagnostic uses. According to Dr Craig Karson, a psychiatrist in Little Rock, Arkansas, eye blinks will soon help diagnose brain disorders, anxiety, and depression just as they now help to diagnose schizophrenia and Parkinson's disease. For many years it has been observed that the patient commonly shows irregularities of eye blinking in both schizophrenia and Parkinson's. But it is only recently that these patterns of irregularities have been understood.

In any event, research on eye blinking is worth keeping an eye on – if for no other reason than to read the moods of others. As Dr Teece adds, 'if your current flame has eye blink storms when professing love, watch out'.

¶ What causes Baggy Eyes?

There are actually rings around the eyes as well as bags under them but the bags look worse and that's why we worry about them more. According to Dr Bette Albert of the School of Public Health at the University of California at Berkeley, bags and rings have the same cause. They are the result of a body fluid problem appearing as puffiness. 'Bags' may occur when fluid accumulates in the area under the eyes. This is where the skin is thinner than anywhere else on the body. With advancing age, and perhaps with the assistance of heredity factors, puffiness may become more prominent or even permanent. This is because skin gradually loses its elasticity and may begin to sag.[13]

However, other factors may be involved as well. In some individuals with permanent bags, they may be due to a specific hereditary condition in which the fat that cushions the eyeball protrudes through weakened muscles. Certain medications such as cortisone and allergic reactions to cosmetics, tobacco smoke, or air pollution may aggravate the situation.

Generally, eye fatigue or irritation may make eyes puffy. Thyroid, kidney, or heart disease can also increase fluid retention. This can be particularly noticeable around the eyes. And if this were not enough, another factor in the baggy eye picture is simply the force of gravity. When one sleeps, especially on the stomach, extra fluid may pool in the upper and lower eyelids.

Besides avoiding the factors already mentioned that worsen the problem, there is not much one can do about the puffiness. Sleeping with the head elevated on an extra pillow may be enough to allow gravity to drain the eye area. In severe cases, the sagging tissue or excess fat under the eyes can be removed surgically. The operation is frequently performed on an outpatient basis.

¶ What causes dark circles under the eyes?

Dr Albert also accounts for this phenomenon with much the same explanation. Again, age is a major factor. It seems that dark circles under the eyes also tend to be a family trait and tend to definitely worsen with age. However, they are seldom a symptom of any underlying medical problem.

What appears as a dark or blue-black tint is actually the blood passing through veins which are located just below the surface of the skin. Furthermore, these circles may be darker when the eyes are tired. Dark circles under the eyes often occur in women during menstruation or during pregnancy.

Cosmetic concealers will cover up the problem if regular make-up is insufficient.[14]

¶ Why do I squint?

The glib answer is that we squint in order to improve our vision. So the real question is, why does squinting improve vision?

Without going into a far too technical description, the eye receives rays of light and bends them so that an image is resolved on a small point of the retina. But things can go wrong. If the rays focus in front of the retina, the person has near-sightedness (myopia) and suffers blurred vision of distant objects. But if the rays focus at a point behind the retina, the person has blurred vision of nearby objects (hypermetropia).

According to Dr Stephen Miller, director of the clinical care centre of the American Optometric Association in St Louis, 'the shape of the eyeball and the focusing power of the lens and cornea help determine focus, but the angle at which light rays hit the eye plays a role'. He adds, 'light comes into the eye from all directions'. Furthermore, 'rays entering the eye at an angle from above or below would tend to focus somewhere before or behind the centre of vision' and 'those rays coming in essentially perpendicular to the eye, on the other hand, would tend to be focused more directly on the retina, providing a clearer image of what one is looking at'.

Therefore, according to Dr Miller, 'the basic impact of squinting is to reduce the number of superficial or peripheral rays of light that enter the eye, so only the rays coming directly in are focused on the retina'. Moreover, 'this cuts out a lot of the rays that are out of focus and eliminates a lot of what would otherwise be a blurred image'.

Dr Miller observes that 'one isn't going to solve vision problems by squinting your way through life, but squinting might help someone who has lost his glasses and needs to see a road sign'.

Finally, too much squinting can result in headaches and 'squint lines' in the face. It is probably a good idea to see an eye care professional if you find yourself squinting much more often than before. Sometimes others will point this out to you.

Rather than continued squinting, vision correction is far more preferable in the long run – taking a far-sighted view of course.

¶ What causes a sty?

A simple answer is provided by Dr Bette Albert again. Dr Albert observes that a sty is bascially 'a pimple on your eyelid'. It is caused by an infection in the follicle from which an eyelash grows or in a connected sebaceous (oil-producing) gland. In order to combat this infection, the body musters an additional supply of blood to the area. This brings the body's infection fighters into the front lines of the battle, but it also produces a sty – 'a red painful inflammation resembling a small boil'.

Dr Albert adds that 'in most cases a sty doesn't require special treatment. Within a few days it will come to a head and burst; the inflammation should subside in a week or so. You can hasten the process – and relieve some of the pain – by applying warm compresses for 15 minutes three or four times each day.' However, Dr Albert is quick to caution that a sty should never be squeezed, nor should the eyes be rubbed, since either can spread the infection. Instead, it is recommended that 'once the pus has come to a head, you can speed up healing by carefully pulling out the lash from the infected follicle'.

Moreover, Dr Albert observes, 'an occasional sty is one of the more common eyelid problems'. Nevertheless, 'if you get sties frequently, however, or if they are severe and resist healing, see your doctor. If one is very large, your doctor may prescribe a medicated ointment.'

¶ When a near-sighted person holds a mirror up close, why are distant objects still blurry?

Surprise, surprise about this one! The viewer is not actually seeing the distant object as if it were brought up close by the mirror. Dr Clint Hatchett, an astronomer with the American Museum–Hayden Planetarium in New York, says that 'what you're doing with a flat mirror is that the person is in a sense seeing what is called a virtual image of an object that is visually the same distance away on the other side of the mirror'. That is, 'it is as if he were looking right through the mirror' at the object. He goes on, 'similarly, with a camera, if you are photographing yourself in a mirror, you have to focus for the distance that is twice as far away as the mirror is'. Dr Hatchett has tested this himself by focusing for the distance of the mirror. 'And it came out fuzzy,' he says.

¶ What causes eye colour?

The pigment of the iris gives the eye its colour. This pigment is called melanin. The amount of melanin determines the eye colour. Large amounts of melanin result in darker eyes (black, brown or hazel). Smaller amounts produce lighter eyes (green or blue). People suffering from albinism have pink coloured eyes due to the absence of melanin. Since the iris is transparent, without melanin, the blood vessels of the eyes show through.

The amount of melanin is determined by genetics. The reason that there are more dark-eyed rather than light-eyed people throughout the world is due to the fact that there is a genetic trait-dominance towards more rather than less melanin. This is why when one parent has dark eyes while the other has light eyes, their offspring are more likely to be dark-eyed.

¶ Why does eye colour seem to change with age?

Eye colour changing with age is largely an illusion. It is not so much a question of the eye changing colour, but more a question of the eye changing size – at least part of the eye at any rate. In fact, the colour of the iris does not change with age. However, the pupil (the black hole at the centre of the iris) may shrink with age. Because it is less dark at the centre of the eye, when the entire eye is observed it appears to be lighter in colour overall. Commenting on this illusion, Dr Richard Thoft, a professor of ophthalmology at the University of Pittsburgh, says that, for example, bright blue eyes may slowly seem to turn lighter as a person ages.

¶ Why does the eye have a 'blind spot'?

Each eye possesses a blind spot in the retina which corresponds to the head of the optic nerve. However, no vision is lost despite these visual black holes. An old theory attempting to explain this phenomenon is that the brain simply ignores the visual blind spots. But laboratory experiments showed that this explanation was inadequate when scientists created artificial blind spots in the visual fields of subjects. Research by Dr Vilayanur Ramachandran of the University of California at San Diego and Dr Richard Gregory of the University of Bristol suggests another view. Their theory is that the brain compensates for these natural holes in our visual field by creating a physiological representation of visual information surrounding blind spots. This creation automatically paints a coherent scene by filling in the optical void.[15]

¶ Why do I often look up while thinking?

It is not widely known, but most psychologists believe that people use either vision, hearing or touching when they seek a mental solution to something. An image, sound or feeling from the past, or a new one we construct, aids our thinking. For example, when someone tries to recall the number of degrees in the angle formed by the hands of a clock at 10 o'clock, they might first try to picture the visual image of the face of a clock, then focus in on the 10 and the 12, and then focus in on the big hand pointing to the 12 and the little hand to the 10 (30 degrees). In any case, a visual image is utilised which is based on the sense of sight.

Specially trained psychologists, known as neuro-linguists, have theorised that specific eye movements indicate what sense is being relied upon in any given thought. In fact, there are seven of these. When the eyes move:

'up-right' indicates visually remembered images;
'up-left' indicates visually constructing new images;
'straight-right' indicates auditory remembered
     sounds or words;
'straight-left' indicates auditory constructed new
     sounds or words;
'down-right' indicates auditory sounds or words;

'down-left' indicates kinesthetic feelings which can include smell and taste;

'straight ahead' indicates that information is being accessed.

If this theory is correct, we sometimes look up when we think to draw upon either an old or a new visual image to help us out.

¶ Can eating carrots improve my eyesight?

As enduring as this notion is, there may be some truth in it. This notion is both one of the oldest of old wives' tales and also one of the standards in the 'Mum's useless advice' category. But Mum might be right about eating lots of carrots to improve vision.

According to Dr Dennis Baylor, a professor of neurobiology at Stanford University, carrots are high in vitamin A. Vitamin A is known to be essential for eyesight. A serious deficiency in vitamin A can cause night blindness – a condition known to improve when sufferers eat carrots. Nevertheless, if you are already getting enough vitamin A in your diet, eating carrots probably will have no effect on your eyesight.[16]

Vitamin A is an important ingredient in the body's manufacture of retinal. Retinal is the light-sensitive chemical that enables cells in the eye's retina to react to light. When light strikes a molecule of retinal, it changes the molecule's shape. This activates a cascade of chemical events that result in sending a message to the brain informing it that light has struck the retina. Dr Baylor adds that the eye has two kinds of light-sensitive cells – rods and cones. Rods are the cells we rely on to see in dim light. They are most affected by a vitamin A deficiency. In order to function properly, the rods require that every retinal molecule be in place and ready to respond to light.

A vitamin A deficiency can cause a retinal shortage in the rods. With some of the retinal sites empty, the rods will never fully adapt to the dark. Hence, you will not be able to see well at night. Dr Baylor believes that eating more carrots, or any other source of vitamin A, should alleviate the retinal shortage and restore night vision.

It's not just carrots that can help vision – spinach may do it too. Although there's no guarantee of attaining Popeye's strength by eating spinach, a diet rich in it can help ward off the leading eye disease of the elderly – geriatric macular degeneration. The macula is the part of the retina located at the back of the eye. In geriatric macular degeneration, the macula becomes progressively damaged. As months and years go by, the vision gets worse. Eventually all vision is lost. There is no way to prevent it and no way to cure it.

According to a 1994 study by Dr Johanna Seddon and colleagues from the Massachusetts Eye and Ear Infirmary in Boston, a diet high in spinach results in a 43 per cent lower risk of developing geriatric macular degeneration.[17]

Popeye was right, but for a different reason. And if Mum is getting older, tell her about the spinach findings to help her vision. While you're at it, thank her for her carrot advice. Tell her you're wearing clean underwear too and really make her happy.

¶ Can the eye suffer a stroke?

We all know that the brain can suffer a stroke, but what
about the eye? Although the eye doesn't suffer the stroke,
the optic nerve certainly can – and frequently does.
When the optic nerve suffers a stroke, there is an abrupt
loss of vision.[18]

¶ Will TV cause eye damage?

It is a common misconception that television causes vision damage. Although the contrast between a bright television screen and a dark room will temporarily tire the eyes, there is no long-term eye damage. Furthermore, there is no risk of eye damage from the reflective glare off the screen from a poorly placed lamp or other light source. Nor is there any need to fear that sitting nose-to-screen will cause nearsightedness or damage children's eyes in some other way. According to Dr Theordore Lawwill, spokesperson for the American Academy of Ophthalmology, 'some people with mild cataracts may even see the screen better in dim light'.[19]

Dr Lawwill adds that children 'like to be as close to the action as possible and would climb into the TV if they could'. Nevertheless, young children are able to focus sharply on objects as close as a few centimetres away from their eyes. This distance lengthens as they get older. Thus, Dr Lawwill claims, they 'may block the screen for other viewers, but it won't hurt anybody's vision'.

Dr William Beckner, senior staff scientist at the US National Council on Radiation Protection and Measurement in Washington, DC, also dispels the notion that TV causes radiation damage to eyes. Dr Beckner says that, compared with television sets built 25 years ago when his organisation first warned of possible radiation risks, 'modern receivers are built differently, using lower voltages and better shielding. No matter how close you sit to the set, x-rays just aren't a problem.'[20]

¶ Is there really an 'evil eye'?

Belief in the power of the 'evil eye' to injure or kill still survives today. When one believes in the 'evil eye', one is convinced that someone can physically harm someone else by merely looking at them in a strange way.

The evil-eye belief, in various forms, has been found in most but not all parts of the world. In many places, the belief survives unchanged – as it has for centuries. Modern English expressions such as 'dirty look', 'looking daggers', 'to stare someone down', and of course, 'if looks could kill!', are remnants of belief in the evil eye.

The evil eye has been rarely studied in behavioural science. But what little research does exist shows that we possibly could be wrong in dismissing belief in the evil eye as belonging only to the superstitious traditions of others. In *The Mirror of Medusa*,[21] Tobin Siebers writes that Stanford University students were once studied in an attempt to assess 'the impact of staring in interpersonal psychology'. Siebers adds, 'the students surveyed, although educated and rational individuals, believed that the intensity of the eye could be detected. Of 1,300 students, 84 per cent of women and 73 per cent of men believed that a stare could be felt. Like many modern individuals, these students would have never taken the evil-eye superstition seriously, yet they unwittingly voiced one of its fundamental premises.'

We should not make too much of the Stanford survey. Nevertheless, we just might believe in the evil eye, or at least in one aspect of it, a little more than we think.

Historically, belief in the evil eye was widespread throughout Europe, the Middle East, India and among the lesser known cultures studied by anthropologists. Interestingly, it was unknown among Australian Aboriginals, according to Clarence Maloney in *The Evil Eye*.[22]

Virgil speaks of the evil eye making cattle lean. From its Latin name fascinum we get the word 'fascination'. In ancient Rome, professional 'evil-eye sorcerers' were hired to bewitch enemies. By the Middle Ages, Europeans were so fearful of falling under the influence of the evil eye that persons accused of looking oddly at others were often burned at the stake as witches. In the sixteenth and seventeenth centuries, hundreds of women were killed in this way. Evidence against them consisted solely of accusations that someone had died or fallen ill after 'being looked upon'. During such witchcraft trials, even judges were often so afraid of the evil eye that the accused was led into the courtroom backwards and made to face the wall. Crossed eyes (strabismus), any eye injury causing misalignment, or even a serious case of cataracts could lead to death.

It has been asserted by some historians that the practice of blindfolding a prisoner at their execution was, at least in part, intended to protect witnesses from the condemned person's attempts to seek revenge through use of the evil eye.

There are several theories accounting for the origins and predominance of the belief in the evil eye. One suggests that it possibly comes from the primal fear of early humans that being stared at by a predator foreshadows an imminent attack. The odd stares and angry glances of hostile neighbours, evil spirits, or jealous gods may often have warned of ill happenings to come.

Another theory holds that early humans must have found it frightening to glimpse their own image in miniature reflected in the eyes of others. Thus, they may have taken this as a sign of being in immediate personal danger. Perhaps the fear would be that one's likeness might become lodged permanently within the eyes of another. Thus, a person with the evil eye was capable of stealing the soul. Anthropologists, missionaries, travellers and others have often remarked that, even into this century, in parts of Africa being photographed was believed to result in the permanent loss of one's soul – 'captured on film' as it were.

Supposed remedies for the evil eye are just as curious as the belief in the evil eye itself. The Congolese gave an evil-eyed male sorcerer beer and tobacco to placate him. The Scots tied red ribbons to the tails of livestock.

The ancient Egyptians had a curious antidote for the

evil eye. This was kohl – history's first mascara. Worn by both men and women, kohl was applied in a circle or oval around the eyes. The basic chemical ingredient for this was the metallic substance, antimony. Could they afford it, aristocratic Egyptians had a soothsayer prepare the compound and apply it. However, women often added their own special, usually secret ingredients to the antimony formula. These concoctions were said to convey additional powers to the wearer.

Besides warding off the evil eye, ancient Egyptian mascara may have had an additional justification for use. Darkly painted circles around the eyes absorb sunlight and consequently minimise reflected glare into the eyes. The ancient Egyptians, living in the hot, harsh, sun-drenched climate of North Africa, may have discovered this fact.

But the ancient Egyptians failed to discover another anti-glare device – sunglasses. The earliest sunglasses were devised by the fifteenth-century Chinese. And in southern China at this time, one of their uses was to hide the wearer's expression in court – and to guard against the evil eye.[23]

¶ What is colour blindness?

Total colour blindness (TCB) is rare. This is according to Dr James C. Trautmann, an ophthalmologist at the Mayo Clinic in Minneapolis. Total colour blindness occurs only when a person sees everything in shades of grey. More often, a person with poor colour vision has difficulty seeing reds and greens. The affected person perceives colours as mostly blues and yellows. Yet this individual may learn to perceive red and green by recognising varying amounts of brightness.

Dr Trautmann adds that, as an inherited condition, colour blindness is much more common in boys than in girls. While fewer than one per cent of girls are colour-blind, the figure soars to eight per cent for boys.

Furthermore, colour vision can also be negatively affected by eye diseases such as cataracts, optic neuritis or retinal disease. But Dr Trautmann claims that these cases account for very few instances of colour blindness.

Although inherited colour blindness cannot be treated or cured, it is not a serious medical problem either, according to Dr Trautmann. If a person knows they are colour-blind, they can easily compensate for their condition. For example, when purchasing clothes, they may ask for some assistance in colour co-ordinating.[24]

5 · **The Nose, Ears and Mouth**

Julius Caesar claimed Cleopatra's large one was the key to her beauty. Ol' Juli baby supposedly had a very big one too. So did Leonardo, Galileo, Voltaire and Jimmy Durante. But Edmond Rostand's Cyrano takes the prize. It's a sign of great dignity provided you don't look down it at others too often, keep it out of other people's business, and don't get it out of joint.

Of course, we're talking about the nose here – and hopefully not talking through it. Indeed, the nose is the stuff from which legends are made. And it doesn't have to be a stuffy nose either. Our nose is always a little bit ahead of us. It's as plain as the nose on your face.

But while we're at it, let's not forget the ears and mouth. What was Gable's fame was Dumbo's triumph. And while comedian Joe E. Brown reputedly had the largest one in Hollywood history, for many others we'll never know – their lips are sealed.

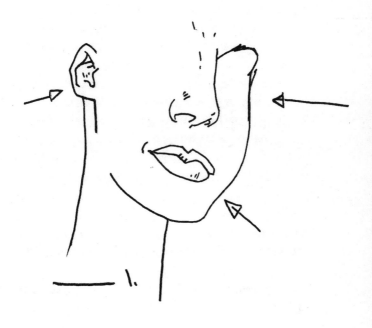

¶ Can my nose grow bigger?

You remember the story of Pinocchio – the little wooden puppet who wanted so much to be a boy. Every time he told a lie his nose grew bigger. In fact the Pinocchio effect is very real, at least to some extent. The inside of your nose is classified as erectile tissue. It's just like your you-know-what or your you-know-what-else. Your nose gets slightly larger or smaller depending on blood flow. What you eat, temperature, illnesses, allergies, even emotional states such as level of anger can alter the size of your nose. It is common for people to feel they have a temporarily stuffy nose after eating. Some even say they can feel it expanding or contracting. The size difference is tiny. Yet if you measure your nose throughout the day, you'd find that it does get smaller or bigger – but never as dramatic as poor Pinocchio's prominent proboscis.

¶ How do people who are totally deaf from birth articulate their thoughts internally?

There have been arguments of all kinds over the need for language in thought and vice versa. The debate probably began with Aristotle who argued that deaf people would never learn language or think. Believe it or not, some people still hold this view. According to Dr Howard Busby, a linguist and speech pathologist at Gallaudet College (for the hearing impaired) in Washington, DC, 'Some people still think that deaf people don't think. But deaf people do think and articulate their thoughts with a variety of languages ranging from English to American Sign Language.'

Dr Busby adds that 'Most people visualise, in their mind's eye, so to speak. What actually makes them think that they're thinking in words is when they try to articulate it. The description of those thoughts gives language to them. The language you choose to describe them forms the shapes of those thoughts. If you articulate them in English, you assume that you thought them in English. 'You can illustrate it this way', as Dr Busby elaborates. 'Think of a high jumper. He's standing back, he's looking at where he's going to jump. He doesn't tell himself, 'I'm going to run this way, I'm going to stop right here, I'm going to spring on my right foot.' He doesn't really say it. He pictures it instead.'[1]

Dr Busby has a reasonable view, but many in psycho-linguistics will no doubt dispute this.

Has anyone asked Aristotle lately?

¶ Why do I yawn?

Yawning is hardly a ho-hum topic. Research reveals there is little scientific evidence to back up many of our popular opinions about why we yawn, when we yawn, what function yawning serves, and what circumstances affect changes in yawning behaviour.

Evidence suggests that yawning is triggered by as yet unknown physiological states. However, it is true that witnessing yawning can immediately provoke yawning, one of the few human behaviours where this occurs. In fact, yawning can be triggered by merely reading about or just thinking about yawning.

Such findings are the product of research conducted by the world's foremost authority on human yawning, Dr Robert Provine, professor of psychology at the University of Maryland. Dr Provine and two other researchers have summarised what science has revealed about yawning so far.[2]

Yawning is a common and probably universal human behaviour. It is performed throughout life. In yawning there is a 'gaping of the mouth accompanied by a long inspiration followed by a shorter expiration'.

Yawning is important in opening the eustachian tubes (which run from the ears to the throat) and in adjusting the air pressure in the middle ear.

Yawning is of clinical importance in health. Yawning,

or its absence, can be a symptom of brain lesions, tumours, haemorrhage, motion sickness, chorea, and encephalitis. It is also an important therapeutic factor in preventing post-operative respiratory complications. It has been reported that psychotics rarely yawn, except in the case of brain damage. Some clinicians claim that those with acute physical illnesses do not yawn until they are on the road to recovery.

Yawning is commonly associated with drowsiness, boredom, and low levels of arousal. Studies confirm, for example, that humans are more likely to yawn when participating in lengthy, uninteresting or repetitive tasks and yawn more when observing uninteresting rather than interesting phenomena.[3]

Dr Provine and colleagues believe that yawning is grossly under-researched. They add that 'beliefs about the relation between sleepiness and yawning are based upon folk wisdom and everyday observations to which science has had little to add'. For instance, texts on sleep only occasionally mention the prominence of yawning in drowsy people, but typically cite only unrelated or general references on yawning.

Neurological evidence for an association between yawning and stretching comes from case reports of brain damaged individuals who are unable to separate the two behaviours. During yawns, such persons often perform related stretching movements of otherwise paralysed body parts. Studies show that drugs that produce yawning also produce stretching in a variety of animals.

Interestingly, Dr Provine's studies show that 'there is also evidence for at least partial autonomy 'of yawning and stretching. Based on one of his laboratory experiments, he adds, 'whereas 47 per cent of stretches were accompanied by a yawn, only 11 per cent of yawns were accompanied by a stretch'.

Furthermore, he theorises that yawning shortly before sleeping and after waking may be either a mechanism to increase alertness or brain function in a drowsy person, or a mechanism to depress alertness, encourage relaxation, or hasten or otherwise prepare us for sleep.

Dr Provine and his colleagues note that 'only a few hypotheses about yawn function have been evaluated'. But no support has been found for the popular assumptions that yawning is either a response to, or somehow regulates, blood levels of carbon dioxide or oxygen. They have found that the yawning rate is neither facilitated nor depressed by breathing gases with elevated levels of carbon dioxide or oxygen. The researchers report that yawning is also unaffected by vigorous exercise.

It has also been found that yawners who yawn infrequently do not compensate by performing yawns of longer duration. Nor do frequent yawners perform shorter yawns.

And there is one thing more. If Dr Provine and

colleagues are correct in all points mentioned above, you have probably yawned at least once while reading the last several paragraphs.[4]

¶ Why do my eyes and mouth sometimes water when I yawn?

The watering of the eyes may result from pressure on the main tear glands, located at the outer margins of the eye sockets, due to the facial contortions involved in yawning. The involuntary act of yawning usually includes opening the mouth very wide while slowly taking in a deep breath.

These same contortions might also put pressure on the salivary glands, especially in a stifled yawn, when the yawner struggles to keep the mouth closed while opening the throat widely. There are three pairs of salivary glands. They are located over the angle of the jaws, in the floor of the front of the mouth, and towards the back of the mouth close to the sides of the jaws.

¶ Why do people hear my voice differently than I do?

This is a classic OBQ. People have been wondering about this oddity since tape recorders were first invented. When you hear your voice played back on a tape recorder for the first time, you can't believe your ears. 'I don't sound like that do I?' No-one can convince you that you do.

Surprisingly, in one sense you're right. It has been discovered that questioning, disputing or completely rejecting a recorded voice as one's own is a worldwide occurrence. Anthropologists have been reporting this for years. They are often the first to introduce tape recorders to the diverse cultures throughout the world as a tool used in their linguistic analyses.

In fact, there is a simple explanation for this curious and seemingly universal auditory phenomenon. According to Dr Nelson Vaughan, a retired speech therapist, diction instructor and voice coach in Hollywood, when we listen to our own voice while we speak, we are not hearing solely with our ears. We are also 'internally hearing a mostly liquid transmission through a series of bodily organs'.

Speech begins at the larynx from which a sound vibration emanates. Part of this vibration is conducted through the air. This part is what others hear when we speak (and what tape recorders record). But another part of the vibration is directed through the various fluids and solids of our head. Our inner and middle ears are located within caverns hollowed out of bone. In fact, this is the hardest portion of the human skull. The inner ear contains fluid and the middle ear contains air, with both constantly pressing against each other. The larynx is also encased in soft tissue full of liquid.

Sound transmits differently through air than through solids and liquids. This difference accounts for nearly all of the tonal variations we hear compared to what others hear.

Interestingly, Dr Vaughan points out that research has found that even young children experience this auditory curiosity. Often they will completely fail to recognise their own voice when it is played back to them even if the playback occurs immediately after they spoke.

Research also fails to show that the voice we hear (our 'internal' voice) is either necessarily pitched higher or lower than the voice others hear (our 'external' voice). Therefore, it could be said that we all really have two voices. Either one is equally our 'real' voice.

It's all in the ear of the beholder.[5]

¶ Why do my ears ring after hearing a loud noise?

This common phenomenon is actually a harmless medical condition known as temporary tinnitus. It is a symptom, not a disease itself. But it can also be chronic and have serious health consequences. In extreme cases, chronic tinnitus sufferers have been known to commit suicide in order to escape the maddening 'ringing'.

Normally, sound comes from a source outside our own body. The sound stimulates the ear's auditory nerve and the brain interprets these impulses as noise. Through learning, we come to distinguish types of noises and judge which are significant and which are insignificant. Strangely, a person with tinnitus receives auditory nerve stimulation from something other than an external source.

Theoretically, virtually anything that can disturb the auditory nerve can cause tinnitus – including extremely loud noises. For example, when a cannon is fired right next to where one is standing, the auditory nerve may remain 'irritated' for several seconds after the shot is fired and the sound ceases. Nevertheless, the 'irritated' auditory nerve still sends impulses to the brain. These are interpreted by the brain wrongly as noise. Hence we experience the ringing until the auditory nerve is no longer irritated. In this case, the tinnitus is only temporary.[6]

According to Dr Tuan Pham, an ear, nose and throat specialist at the Royal Prince Alfred Hospital in Sydney, tinnitus 'is very interesting' and 'needs to be further investigated'. Dr Pham points out that, besides the ringing sensation, tinnitus sufferers can hear 'hissing', 'buzzing', 'humming', 'a waterfall-like sound', or 'a pulsatile sound' (corresponding to one's own heartbeat).

He adds that, besides being a reaction to a loud noise, temporary tinnitus is most commonly caused by vascular distress after a physical or mental trauma or by an allergic response to medication. Fluid tablets or aspirin may be involved as well. In one study he cites, people who took 20 aspirins per day and were subject to tinnitus attacks afterwards saw these attacks completely disappear when they discontinued the large aspirin doses.

The causes of chronic tinnitus are numerous: inflammation of the ear, ear canal problems, jaw problems, work noise exposure, old age, clogging of the ear by earwax, vertigo attacks, excessive use of the telephone, muscle spasms in the ear, nutritional deficiencies, allergies and infections, among other factors.[7]

Chronic tinnitus sufferers not only have to live with the infernal noises, they often experience hearing loss as well.[8]

According to Dr Pham, there are various measures to help a patient cope with chronic tinnitus. Chief among these is 'reassurance'.[9]

¶ Can I permanently change the sound of my voice?

You can definitely change the sound of your voice. A US speech pathologist claims that 30 per cent of people are unhappy with the sound of their voice. Dr Daniel R. Boone of the Department of Speech and Hearing Science at the University of Arizona in Tucson, adds that 'older people want to sound younger, and the young want to sound older'. And with a lot of business conducted over the phone, your 'voice is becoming as important as your looks'.

'We live in a society where we want to sound a certain way,' says Dr Morton Cooper. Dr Cooper, a Los Angeles speech pathologist, adds that 'we want our voice to sound not as nasally, but rich and full. We want authority in our voice. We want to sound as if we have status and position.'[10]

Dr Lillian Glass, a speech pathologist in Beverly Hills, California, is known as 'speech consultant to the stars'. Her clients have included Dustin Hoffman and Sean Connery. She claims that 'often you can give a false impression by the way you speak. A rough, harsh voice can give the impression that a person is mean and gruff. A high-pitched sound can signal immaturity. If you have a monotonous voice, people may think you're not interesting – or interested.'[11]

Just like cosmetic surgery, voice improvement is becoming more and more fashionable. In the US, professional speech consultants are doing a booming business. It seems that any voice can be improved. Speed, tone, degree of nasality and even accent can be changed. According to Dr Boone, 'sometimes all it takes is just training people to speak a little faster'. In the US, where the southern accent is seen as 'lower class', many individuals from the south seek to obtain a different accent. As University of California–Berkeley educational psychologist Dr Lawrence Stewart remarks, 'when people hear a southern drawl, they automatically deduct 20 IQ points'.

Dr Glass adds that six months of voice training can work wonders since 'nobody is born with a bad voice. Anybody can improve the way they speak.'[12]

¶ Can illness cause a change in accent?

Sometimes an accent can completely baffle the experts. A US doctor has reported the strange case of a thirty-two-year-old Baltimore, Maryland, man who suddenly began speaking with a Scandinavian accent. The man displayed a rare disorder that may shed light on not only how the brain recovers from a stroke, but also how the brain produces language.

The man suffered from what is termed 'Foreign Accent Syndrome' (FAS). As a monolingual American-English speaker, he had no previous experience whatsoever with any foreign language. Nothing in his past pointed to fluency or even familiarity with any foreign language. Yet when he spoke after suffering a stroke, he sounded both Nordic and totally unfamiliar with English, according to Dr Dean Tippett.

Dr Tippett, a neurophysiologist at the University of Maryland School of Medicine in Baltimore, added that 'he was pretty clear, everyone who heard him said he sounded Scandinavian or Nordic'.

Foreign Accent Syndrome is a neurological condition in which a brain malfunction produces speech alterations which sound like a foreign accent. The syndrome is triggered by stroke or head trauma. There are reports that FAS has produced German, Spanish, Welsh, Scottish, Irish and Italian accents.

In a 1990 paper to an American Neurological Association meeting in Atlanta, Dr Tippett argued that studying the intricacies of FAS may reveal secrets about how particular parts of the brain contribute to spoken language. Monitoring the course of the return of a patient's original accent may be a useful means of tracing the process of stroke recovery as well.

In the case of this patient, immediately after the stroke the man's speech was slurred for a day or two. His Scandinavian accent appeared only as he started to recover. He added extra vowel sounds as he spoke, saying such things as, 'How are you today-ah?' His voice also rose in pitch at the end of sentences, as if asking a question. Moreover, 'that' was pronounced as 'dat'. Some vowel sounds were substituted as well, making 'hill' come out as 'heel' and 'quite' come out as 'quiet'. Often the vowel sound was exaggerated and drawn out.

Perhaps the world's foremost authority on FAS is Dr Arnold Aronson. Dr Aronson is a speech pathologist at the Mayo Clinic in Minneapolis. He has evaluated about 20 people with the syndrome. According to Dr Aronson, non-US cases have produced a French accent in a British person and a Polish accent in a young Czech. Interestingly, about 40 per cent of cases produced German, Swedish or Norwegian accents.

Intriguingly, the trauma-acquired foreign accent may become 'rather permanent' depending on where the brain is injured, according to another authority, Dr Elliott Ross. Dr Ross is the director of the clinical research program at the Neuropsychiatric Research Institute in Fargo, North Dakota. In the case of Dr Tippett's patient, the man's speech returned to normal about four months after the stroke.

Dr Tippett reported that his patient suffered no additional pain in speaking differently. Indeed, at first the man even 'enjoyed his new accent, saying he hoped it would help attract women'. But by the time his accent had almost completely faded, he said he was happy to be speaking like an American again.

¶ What causes 'varicose nose'?

What do legendary comedian W.C. Fields, Santa's reindeer Rudolph and thousands of people throughout the UK have in common? Their bright red noses of course.

Every two years in the UK we have Comic Relief Day, otherwise known as National Red Nose Day. Many of the famous and not-so-famous will pay for the privilege of wearing a plastic red nose to raise money for poor and disadvantaged people in the UK and Africa.

But what of the red nose condition itself? What is it called? What causes it? How can it be prevented or treated?

Contrary to popular opinion, the red nose condition is not called 'varicose nose'. Nor is it caused by too much alcohol, although alcohol can aggravate the problem. W.C. Fields was known as a Hollywood star who drank heavily. His alcoholism and 'varicose nose' became associated, but really there's no causal connection. Instead, the condition is actually a form of acne called rosacea.[13]

In the *Journal of the American Medical Association*, Marsha Goldsmith explores developments in the treatment of rosacea.[14] For years, doctors have treated this condition with antibiotics, often without success. But a few years ago, doctors at Harvard University found that a cream commonly used to treat (of all things) vaginal infections is much more effective.

The doctors tested the cream, which contains metronidazole, on 40 men and women with moderate to severe rosacea. After three weeks, the cream had cleared up half the lesions and redness and relieved the itching and dryness that accompany rosacea. Among those who used a placebo cream instead, there were no noticeable changes. The doctors are not sure why the cream works, but they think that perhaps metronidazole kills a mite called Demodex folliculorum that burrows into the pores of the skin. The presence of too many of these mites can inflame the skin.

Dr Jonathan Wilkin of Ohio State University's Department of Dermatology says that 'rosacea is truly a cutaneous disorder', not just 'a complexion problem that runs in my family', as many people think. It can be treated easily now and reversed; if not treated, it will probably grow worse.'

Every two years on National Red Nose Day we will be putting on our red noses for one day for a good cause. But for others, they may one day remove their red noses for good.[15]

¶ What is a hiccup?

A hiccup is actually an irritation of the diaphragm that causes a spasm. In fact, a hiccup is a two-part phenomenon. First, the diaphragm contracts involuntarily. This is because the nerves that control it have become irritated by something. Perhaps it was that we ate or drank something too fast. When breathing and eating have to be undertaken at the same time, you invite irritation. Usually you control your own hiccupping by not doing the things that can provoke them. Second, when air is inhaled, the space between the vocal cords at the back of the throat (the glottis) snaps shut with characteristic clicking sound. This snapping is what we hear when we hiccup.

To understand a hiccup, you have to understand what the diaphragm is. The lungs are enclosed in a kind of covered cage in which the ribs form the walls and the diaphragm forms the floor. The diaphragm is an upwardly arching sheet of muscle. When you take a normal breath, the diaphragm is drawn downwards until it becomes flat. At the same time, the muscles that surround the ribs contract, lifting the lungs up. It is rather like the lifting of a hoop skirt in an updraft. In this way, the chest cavity becomes wider, deeper, and its air capacity increases.

There are many remedies for hiccups. Some people drink a glass of water without pausing for air. Others hold their breath until the hiccups stop. Still others breathe into a bag. These techniques may restore the normal rhythm of the twitching, irritated diaphragm. Perhaps this happens by reducing the oxygen level and increasing the carbon dioxide level. Other cures attempt to trick the nervous system with diversionary tactics such as tickling the nose to induce sneezing or pulling the tongue. If hiccups persist in spite of all efforts to stop them, medical attention should be sought.

The medical name for a hiccup is singultus. Hence, don't sing alone – sing ult us.

There is a medical condition known as epidemic hiccupping. It is a symptom of some forms of encephalitis (inflammation of the brain).

The record for the longest span of uninterrupted hiccupping is (are you ready for this?) 68 years. The unfortunate man hiccupped an average of 20 to 25 times per minute – but otherwise led a normal life, marrying twice and fathering eight children.[16]

¶ What is earwax?

It's sticky, it's ugly, it's yucky, but it's important to health – it's earwax.

Earwax is sticky on purpose. Dust, dirt, bacteria, fungi, and other foreign dangers to the body all stick to the wax and thus do not enter the ear – one of the most sensitive areas of the body and quite exposed when you think about it.

Earwax also contains special enzymes called lysozymes. Lysozymes break down the cell walls of foreign bacteria and are also contained in saliva. So earwax fights bacteria in two ways – sort of acting like flypaper to halt them and then biochemically dissolving them.

The medical name for earwax is cerumen.

Also, there are different colours of earwax among the various peoples of the world. In white and black people earwax is honey-coloured, moist and soft. But in certain Asian groups (Mongolians, for instance) it is grey, dry and brittle. There is a specific gene for earwax. Wet is the dominant trait; dry is the recessive.

Although we are taught to clean our ears of earwax, from a health standpoint it is probably a good idea to leave some of it there. Just as it's wise to be lucky, it's wise to be yucky![17]

¶ How do we swallow?

A normal swallow is a three-phase process taking from eight to 12 seconds from beginning to end. Part of the swallow is voluntary, but most is involuntary. The first phase of food swallowing is the termination of chewing. Chewing is the first and last voluntary aspect of the swallow. Chewed food in the mouth is mixed with saliva to prepare it for passage down the throat (pharynx). The actual swallow begins as the food is pushed to the back of the throat by the tongue. The tongue does this by pressing against the roof of the mouth (soft palate), forcing the food into the throat.

The second phase is when the food is in the throat. This takes about two seconds and is involuntary. These involuntary reflexes are controlled by the swallowing centre located in the brain stem. Food is pushed down into the oesophagus by rhythmical contractions of the oesophageal muscles (peristaltic waves). It is important that the sphincter muscle at the entrance to the oesophagus remains relaxed in order to open the channel of the throat. Also, the larynx must elevate to force the epiglottis to close over the airway (trachea) to prevent food from entering the lungs.

In the final phase, the peristaltic waves continue to push the food through the oesophagus and down into the stomach. One more muscle sphincter must relax for this to occur and also to help prevent food from coming up again. This last phase takes between six to ten seconds.

With such a complicated process, it is no wonder that something can easily go wrong.[18]

¶ Does my heart stop when I sneeze?

This is one of the most commonly held body myths. Your heart does not stop when you sneeze, although it might feel like it does. Instead, what happens when we sneeze is that a rather powerful positive pressure condition is created in the chest. It is fairly violent and can actually change the rhythm of the heartbeat. But it does not stop the heart from beating. This feeling of a momentary 'jump' in the heartbeat is probably the origin of the erroneous belief that the heart stops. According to Dr Jay Block, former president of the American College of Chest Physicians in New York, the creation of positive pressure in the chest when we sneeze or cough has a name – the Valsalva manoeuvre.[19]

¶ Why do I sometimes sneeze when I come out of a dark room and into the daylight?

Even if this doesn't happen to you, just watch people leaving a theatre at the end of a matinee to see that it certainly happens to others.

Francis Bacon (1561–1626) discussed 'light sneezing' in *Sylva Sylvarum* (1635). He erroneously believed that it was caused by moisture being drawn down from the brain to the nostrils and eyes due to a 'Motion of Consent'.[20]

But 'consent' has nothing to do with it. According to Dr R. Eccles of the Common Cold and Nasal Research Centre of Cardiff, Wales, this is termed the 'photic sneeze'. In fact, this is a genetically transmitted characteristic which affects between 18 and 35 per cent of the population. Dr Eccles writes that, 'The sneeze occurs because the protective reflexes of the eyes (in this case on encountering bright light) and nose are closely linked. Likewise, when we sneeze our eyes close and also water. The photic sneeze is well known as a hazard to pilots of combat planes, especially when they turn towards the sun or are exposed to flares from anti-aircraft fire at night.'[21]

6 · **The Skin**

What is the largest human organ? You were right if you said the skin. The skin is also the heaviest organ, weighing between 2.5 and 4.5 kg. If you were to spread it out like a pancake, it would cover an area of about 2 m². It forms a perfect protection for us, provides great insulation against heat and cold, and keeps out bacteria and other things that would harm us. It also serves as the end point for our nerves so that we can touch. In fact, skin is pretty fantastic!

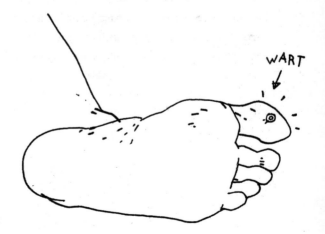

WART

¶ Why do I itch?

Although science tells us much about itching, we still
don't understand everything about what is medically
termed 'pruritis'.

Sensory receptors located just below the surface of
the skin send messages to the brain. Itch sensations flow
along the same pathways of the nervous system as pain
sensations. The vast majority of sensory receptors are
'free' nerve terminals which do not seem to be designed
for any single function. They carry both itching and pain
messages.

These sensory receptors are the most common nerve
terminals in the human body's entire nervous system.
This is understandable since the entire body's surface
must be covered. When these sensory receptors operate
at a high level, a pain signal results. But when only a low
level registers, an itch signal results.

Scientists can induce itching by heating a subject's
skin. However, if too much heat is applied, then pain is
produced instead. Certain chemicals can also bring about
itching in humans during laboratory experiments. Chief
among these are the histamines. This is why doctors
often treat pruritis complaints with anti-histamines.

When a doctor sees a patient who complains of itchy
hives, for example, the doctor treats the hives and the
itchiness stops. But this does not mean that anyone

knows for sure why the itch is associated with the hives. Itchiness can also be associated with a variety of serious illnesses (such as Hodgkin's disease) and can indicate the onset of others (such as diabetes).

The function of itching is open to speculation. Some physiologists maintain that itching provides a first warning to the body of imminent pain. Some anthropologists suggest that, in our primitive past, itching served the purpose of telling early humans when it was time to pluck out body lice and other skin parasites.

Strangely, research shows that itching may be less easy to live with than pain. Doctors note that patients with severe itching are invariably more willing to 'scratch till they bleed'.

Itching has been compared to tickling. However, evidence suggests that these reactions are entirely separate. Ironically, although tickling produces laughing while itching and pain do not, subjects can take only short periods of tickling compared with the other two. In this sense, it is not always true that one chooses pleasure over pain.

It seems there are continually new sources of itching. According to Dr S. Leonard Syme of the School of Public Health at the University of California at Berkeley, there seems to be an epidemic of pruritis ani – anal itching due to over-vigorous wiping with abrasive toilet paper.[1]

Nevertheless, science has much to learn about

itching. As dermatology professor Dr Ronald Marks from the College of Medicine at the University of Wales recently remarked, 'there are many areas with persistent pruritis that we find enormously difficult to handle clinically. Treatment with conventional therapies is hopelessly inadequate.'

One 'itching' myth has definitely been dispelled: there is no evidence for the so-called 'seven-year itch'. However, George Axelrod's famous play and subsequent classic Billy Wilder film by that name – about the roving male libido after seven years of marriage – is a cherished part of our folklore, if not our medicine.

The enigma of itching is irritating – and scientists continue to scratch.[2]

¶ Why do I get chills when I hear chalk screech along a blackboard?

It crawls up and down your spine. You wince, you shiver, you cringe at the mere thought of it. Your worst enemy or perhaps your best friend might have done it to you. The ancient Greek philosopher Aristotle first observed it and remarked on its chilling consequences. Modern lexicographers have resurrected the archaic verb 'gride' to describe the act of causing it. And science is still puzzled by it.

What is it? It is that awful, high-pitched, scraping sound you often heard at school when chalk is dragged across the blackboard – 'blackboard screech'.

Why do we get chills when we hear blackboard screech? The short answer is that we don't really know. Despite the antiquity of its discovery and the near universality of its spine-tingling effects, surprisingly little is actually known about this odd auditory phenomenon.

No scientific studies examined blackboard screech until 1986. Previously, it was commonly believed that the high frequency sounds it possessed were responsible for the unpleasantness experienced by those hearing it. However, when this theory was tested in the laboratory, it was proven to be incorrect.

In 1986, three US researchers at the Cresap Neuroscience Laboratory of Northwestern University near Chicago conducted four experiments to test the 'psycho-acoustics' of blackboard screech.[3] But these experiments actually raised more questions than they answered.

The researchers, Drs D. Lynn Halpern, Randolph Blake and James Hillenbrand, first recreated blackboard screech by dragging a garden rake across a slate surface. They recorded the sound and ran the recording through different frequency processors.

The original theory that blackboard screech produces chills simply because of its high frequency quickly fell by the wayside. When they removed the high frequencies, it had no affect on the unpleasant reactions experienced. When the lower frequencies were removed, surprisingly the sound that was left was judged by subjects in the experiments to be rather pleasing. Next, the researchers experimented with the volume. This had no bearing on the chilling effect. Finally, they compared blackboard screech with a variety of sounds found in nature.

Interestingly, it was discovered that blackboard screech bears a remarkable resemblance to the warning cries emitted by a species of lower order primates: Japanese macaques. Due to this finding, the researchers suggested that our tingling spines may be a primitive reflex left over from an earlier phase of our evolutionary history. Alternatively, blackboard screech may be similar to the sound made by some prehistoric predator.

Yet these are only conjectures.

As the researchers conclude, 'Regardless of this auditory event's original functional significance, the human brain obviously still registers a strong vestigial response to this chilling sound.'[4]

¶ Why does my skin wrinkle?

The conventional answer is that wrinkles are caused mainly by exposure to the sun's ultraviolet radiation. Over time, the ultraviolet rays break down the middle layer of the skin (the dermis) causing it to loosen and allowing wrinkles to form. This condition is known as 'dermatoheliosis' – sun-induced skin aging. Yet, with extreme old age, the dermis loosens of its own accord.

According to Dr Allen Lawrence, a noted dermatologist in Chicago, wrinkles are more likely to occur among people who have thin layers of skin or whose skin is lighter in colour.[5] Exposure to the sun is to be avoided as much as possible to preserve the skin. The next best step is to use a sunscreen.

Aged skin is thinner than younger skin and less cellularly organised. Under the microscope, where healthy basal cells once stood lined up in neat columns within the outer layer of the skin (the epidermis), now there is disarray – a breakdown in the normal process of cellular proliferation and organisation.

With age, collagen fibres decrease in number, organisation, and density. Collagen fibres are important in maintaining skin integrity. And also with age, the formerly smooth, ribbon-like elastin fibres become coarser, denser, and less resilient. Elastin fibres are responsible for the skin's ability to 'snap back' to shape after being stretched, particularly after the skin has been exposed to sunlight.

Other things happen when skin ages too. The tiny blood vessels in the dermis become thick-walled. Yet strangely, they also become leakier. There is a general loss of nerve cells, hair, sweat ducts, and sebaceous glands that produce fatty secretions called sebum.

Skin aging is, therefore, a combination of two processes. One is intrinsic, the other extrinsic. Chronological aging is genetically programmed but the chemical reactions triggered by exposure to sunlight are environmental. We may not be able to do anything about our genes, but we certainly can control our behaviour and stay out of the sun.

The medical term for a wrinkle is (are you ready for this?) 'wrinkle'. The origin of this word is obscure but it probably comes from the Old English word 'gewrinclod' which means 'the winding of a ditch'.[6]

¶ Why do the ends of my fingers and toes wrinkle after a bath?

This is one of the classic OBQs. The short answer is that this form of wrinkling occurs because of differences in the layers of skin and our skin's natural protective oils.

According to Dr Marianne O'Donoghue, Professor of Dermatology at Rush University in Chicago, when we are immersed in a bath, 'the top layer of the skin absorbs a lot of water. The bottom layer of skin can't get any bigger, so the top layer must corrugate or pleat. Fortunately, the effect is reversible.'

But you might ask, why do we shrivel up like a piece of dried fruit and not swell up like a sponge? Without our skin's protective oils, the palms of our hands and the soles of our feet become dehydrated in the presence of water, not hydrated. It works this way.

About 75 per cent of our body is water. This amount varies somewhat depending on the amount of fat we have. Dehydration occurs when the protective oil of our skin surface is washed away. Our body water begins to leak out of our skin cells. These cells have semi-permeable membranes, which means that they can easily give up water but cannot absorb it as well. Once the oil is lost, after about 15 minutes of immersion, the flood gates open outward. The water moves out and wrinkling occurs.

The ends of our fingers and the palms of our hands wrinkle more quickly than the backs of our hands because the backs of our hands have extra sebaceous glands. These glands continually replenish those protective oils almost as fast as they get washed away.[7]

¶ Is it possible to stop wrinkles?

This book is trying to stay away from therapeutic suggestions, but we just can't help ourselves here.

Are those first tiny crow's feet starting to appear at the outside corners of your eyes? And what about under them? Are those forehead lines really caused by smiling? How would you really like to laugh at your first wrinkles – and at all those yet to come?

Recently, a US medical team has discovered a new elixir that stimulates the growth of skin cells. Applied as a cream, it could well prove to be the fountain of youth for aging skin.[8] The elixir is composed of the hormone PTH minus a few amino acids. PTH is produced in certain neck glands. Dr Michael Holick and colleagues at the Section of Endocrinology, Diabetes, and Metabolism at the Boston University School of Medicine injected laboratory rats with the PTH compound. There was a dramatic 300 per cent increase in the rate of skin cell growth. The rats emerged with younger, softer skins.

One of the reasons that skin looks wrinkly with age is that skin cells are not replaced as fast. The PTH compound may also be useful in treating scarring, burns and even baldness.

Hopefully, what'll work in rats will work in humans. After all, Dr Holick agrees, 'the two are pretty similar in lots of other ways'.

¶ What are stretch marks?

According to Dr Alan Xenakis in *Why Doesn't My Funny Bone Make Me Laugh?* (1993), stretch marks are merely evidence that the skin has been forced to expand in order to accommodate a somewhat larger body mass. The skin has an amazing ability to expand and contract as the body mass changes. Stretch marks often occur after dieting.[9]

It is a myth that stretch marks occur more often in women than in men, although it may seem so. One of the reasons for this is that hair is more likely to cover men's bodies rather than women's in those places where stretch marks are most likely to be seen.

Where are those places? You're right about that too.

¶ What are warts?

Warts are infectious swellings or benign tumours in the outer layer of the skin. Because they are contagious, they can spread from person to person. Most frequently, they appear on the hands, fingers, and soles of the feet. However, genital warts can occur on the genital and anal areas.

Common warts are caused by the human papilloma virus (HPV). The virus can remain inactive in the body for six months before warts appear.

The medical term for a wart is 'verruca'. There are at least 26 types of verruca. The common wart is sometimes called the 'seed verruca'. It is so called because of the tiny black specks or 'seeds' within it. In fact, these 'seeds' are actually elongated capillary loops that have thrombosed (clotted). Another interesting verruca is the 'soot verruca'. This is the so called 'chimney-sweeps' cancer' but to call this wart a benign tumour is a misnomer. 'Chimney-sweeps' cancer' is really a malignant carcinoma of the scrotum due to soot poisoning.

It was one of the writer Roald Dahl's jokes that he named one of his characters after warts. This is the obnoxious Veruca Salt ('the girl who got everything she wanted') in *Charlie and the Chocolate Factory* (1964). Why Dahl dropped an 'r' is known only to him. Dahl's subtle humour also extended to Veruca's surname.

Throwing salt on a wart is one of the traditional folk customs for getting rid of them.

As people age, they tend to build up an immunity to warts. Nevertheless, they can occur at any age.

Normally, warts do not themselves endanger health. However, see your doctor about any wart you might have – just to be sure. Genital warts should be seen by a doctor. There is evidence that such warts lead to cervical cancer. Plantar warts (the warts on the soles of the feet) should be seen by a doctor too. These warts can be particularly painful as well.

Studies show that two out of three cases of warts disappear on their own within two years. That is why physicians now usually recommend leaving warts alone. But there are two conflicting theories about the treatment of warts.

One view says that, since warts occur most commonly in children, it is believed that children should not have warts removed in order that the immune system can be boosted by having to cope with the virus.

The other view holds that some warts should be removed. This is because doctors have found that if one or two warts are treated, others on the same person disappear without treatment. It is believed that wart removal may stimulate antibodies to fight the virus.

Again, see your physician about any of this, but over-the-counter treatments often involve salicylic acid.

Cryotherapy (freezing) is often employed and sometimes surgery, including laser surgery. In 1994, a story in *Australian Doctor* described how the drug cimetidine was 'remarkably effective at curing warts in children'. Hypnotherapy has been used to treat warts. It seems to be particularly effective in treating children, according to Dr Robert Noll of the Children's Hospital Medical Center in Cincinnati. Dr Noll writes, 'Nearly all patients (86 per cent) were completely cured of their warts within three months of the onset of therapy.'[10]

The fact that most warts vanish by themselves after two years increases their mystery. It also accounts for why folk beliefs were, and still are, thought to cure warts. The warts go away by themselves and then the folk cure gets the credit. Medical experts agree that the worst thing one can do is cut out the wart yourself. Besides endangering yourself by the use of unsterilised instruments and poor surgical technique, you may re-infect yourself with the virus that caused the wart in the first place.

And that would be like throwing salt onto the wound.[11]

¶ What are genital warts?

Genital warts are somewhat different. According to Dr June Reinish, head of the Alfred Kinsey Institute in Minneapolis, genital warts are also caused by a virus – the human papilloma virus (HPV). They are smallish tumours outside and inside of the genitals. But they do not always look like a wart. Indeed, some think genital warts look rather like cauliflower in colour and appearance. If untreated, genital warts can lead to abnormal cell problems in a woman's cervix. This has been linked to cervical cancer in studies. The genital wart virus has also been linked to penile cancer.[12] Genital warts should be seen by the family physician. There are treatments but they are more difficult to treat in women due to their location.[13]

¶ Can kissing a frog or toad give me warts?

Many folklore solutions exist to rid yourself of warts. For example, one of the more memorable was from the MGM classic film *Huckleberry Finn* (1939), starring Mickey Rooney, which was, of course, based on Mark Twain's literary masterpiece, *The Adventures of Huckleberry Finn* (1884): 'You go to a cemetery where someone wicked has just been buried. You bring a dead cat in a gunny sack. At midnight, when the devil comes to take the soul of the wicked person, you swing the sack around your head three times and say, "Devil take d'wicked, d'wicked take d'cat, d'cat take d'warts, I'm done wid ya." And you throw the sack as far as you can.'

Touching a frog or toad may not give you warts, but kissing or licking one can give you a 'high'. This belief is in our folklore, in our nursery rhymes, and now (surprise, surprise) it's in our science.

Let's start at the beginning. The Frog Prince legend is part of the folklore tradition of most of the world. In fact, it is the first story in *Nursery and Household Tales* (1812) by the Brothers Grimm.

In the most common version of the legend from thirteenth-century Europe, a princess asks a frog to recover her lost ball from a lake in exchange for a trip to her castle. After the frog retrieves her ball, the princess forgets her promise. Nevertheless, the frog unexpectedly shows up in her bedroom where the princess continually rejects his advances. She eventually kisses him, whereupon he turns into a handsome prince. Needless to say, they marry and live happily ever after.

In 1991, two US doctors claimed they discovered why this legend is so widespread – it all has to do with biochemistry.[14] Bufotenin is a chemical that produces hallucinations in humans. It also has aphrodisiac effects, particularly with women. High amounts of bufotenin have been found to be present in the skin of many common frogs and toads. Thus, 'kissing or licking' one of these warty creatures 'can result in vivid hallucinations', romantic impulses and sexual thoughts, according to Drs David Siegel and Susan McDaniel of the School of Medicine and Dentistry at the University of Rochester.

The two doctors add that 'this biological property' was well known to people throughout the world for centuries. And it also explains why frogs and toads are often portrayed in folklore as 'creatures of transformation, or as intermediaries with other worlds'.

Moreover, the two doctors note, 'the magical moment' of the kiss 'is, in fact, an aphorism known to many young women, that one must kiss many frogs to find one's prince'.[15]

¶ What are goosebumps?

Although many claim they get goosebump-type sensations from a wide range of stimuli, a goosebump is actually a muscular reaction to cold. When exposed to low temperature, the small muscles at the base of each hair start to contract. The effect is the formation of a mound around the hair. If the temperature stays low enough for long enough, a goosebump clearly forms and the hair stands erect. The medical name for goosebumps is 'cutis anserina'. Goosebumps are reported from all around the world. It appears that no nation or ethnic group is immune.

Hair functions to protect the body against the harsh rays of the sun. This is probably why we have the most body hair on our head. The head needs the most protection as the brain can be 'cooked' inside the skull if unprotected on a hot, sunny day. This is particularly true if the body undergoes great exertion – as in 'jogger's heat stroke'. Hair also reduces friction problems in body movement. This is perhaps one of the reasons why we have hair in our armpits and between our legs.

But hair also insulates the body against cold. In animals covered with hair, the rising strands of hair form a protective shield against cold. Cold air is trapped in and between hairs instead of coming into direct contact with unprotected, sensitive skin.

Goosebumps may be a carry over from our earlier, more primitive primate existence. Although we humans have lost most of our body hair compared with gorillas, chimpanzees, or orang-utans, we still experience the same muscular contractions in reaction to cold as they do. But since we have less hair than our more primitive brothers and sisters, our goosebumps are more noticeable.

When an animal's hair stands on end, it looks ferocious. When ours stands on end, we just look cold.[16]

¶ Why don't I get goosebumps on the palms of my hands or the soles of my feet?

This may be the shortest answer to an OBQ anywhere in this book. We only get goosebumps where we have hair. This is why we do not get goosebumps on our palms or soles.

¶ Why does my skin burn?

Our skin can burn simply because our bodies can be fuel –
particularly our fat. The next time you roast some meat
over an open flame, watch the drippings fall into the fire.

¶ What is meant by the terms first-, second- and third-degree burns?

Contrary to common belief first-degree burns are not the most serious. The degree of a burn refers to what extent heat has destroyed the layers of skin. A first-degree burn, such as a minor scald or sunburn, affects only the outer skin. These burns heal by themselves in a few days without scarring.

A second-degree burn is somewhat deeper and actually destroys some layers of skin, resulting in the formation of blisters. There is a moist, whitish surface colour. If blisters are unbroken, they protect the injured area. Again, no scarring results and the skin regenerates after a few weeks. According to the staff of the Royal Children's Hospital in Melbourne, a somewhat more serious burn than a second-degree burn is a 'deep second-degree' burn. In this, there is a 'moist white slough, red mottled' surface colour.[17]

In a third-degree burn, heat completely destroys the upper layers of skin including accessory skin structures such as hair and sweat glands. The burn penetrates to the subcutaneous layer. The burned area takes on a dry, charred, white, leather-like appearance and normal skin does not regenerate. Such burns demand immediate medical attention.

Third-degree burns are often treated with skin grafts.

Thin layers of skin are removed from an unburnt part of the body and grafted onto the burned area. The donor site usually heals without problems since only a thin layer of skin is taken. Skin grafts are not performed for merely cosmetic purposes. Skin grafts may be necessary in third-degree burns because exposed subcutaneous tissue cannot heal fast enough to protect the body from infection and from loss of fluids.

Problems arise if the burn site is so extensive that there is not enough unburnt skin on the body to perform sufficient grafts.

Interestingly, the skin used for the graft may be considerably smaller than the area of the burn to be covered. This is because a variety of techniques are available to 'enlarge' the grafted skin. One technique is to cut the skin into filigree threads to cover the burn. Another technique is to mince the skin, place it in a nutrient solution, and grow skin sheets in a laboratory.

Pigskin and skin from cadavers can be employed as temporary grafting sources but the body eventually rejects them.

Recent attempts to develop artificial skin have had mixed results. Among those developed in recent years is one consisting of a blend of substances including shark cartilage. However, most doctors still believe that a graft of the patient's own skin is normally preferable since the body's immune system is more likely to reject any artificial skin as foreign.

Nevertheless, experts in this field believe that a rejection-free and fully tested artificial skin is likely to be available by the end of the century.

As research continues, in many respects it has only scratched the surface.[18]

¶ Why do men have nipples?

In theory, we all could have fully functioning breasts capable of giving milk. But male breasts, including the nipples, do not get enough of the female hormone oestrogen so they never achieve the ability to lactate.

According to an international expert on sex hormones, 'there are exaggerations in each sex based on early development and later hormonal influence. The nipple is one of the features whose full development has been restricted in the male during early development so it never develops its full function, which is found in the female.'

Dr Bruce McEwen, a neuro-endocrinologist at the Rockefeller University in New York, adds that the presence of nipples and other breast tissue in men illustrates the fact that the basic body plan of men and women is similar.[19]

As far as we can tell, male breast tissue has no functional significance other than perhaps cushioning the heart and lungs from injury. However, nipples are considered erogenous in men as well as in women.

Nevertheless, Dr McEwen points out, 'the tissue is still there and can actually respond to female hormones in the male, as exemplified by the phenomenon of gynaecomastia [abnormal enlargement of the breasts in men]'. Gynaecomastia occurs when there is an excess of oestrogen. It is commonly seen in male alcoholics as well.[20]

¶ Could men ever give milk?

With the right hormonal treatment, this is no problem at all. Dr Jared Diamond, professor of physiology at the UCLA School of Medicine, wonders why nature didn't give men the capability to suckle too. Wouldn't that double the number of people a baby could rely on for sustenance? In any case, Dr Diamond writes, 'Experience may tell you that producing milk and nursing youngsters is a job for the female mammal, not the male. But your experience is probably limited, and the potential of biology – and medical technology – is vast.'[21]

¶ Why do we blush?

This uniquely human physiological phenomenon has been fascinating and baffling since the first caveman committed the first social faux pas at the first human meeting place. Researchers say that by studying why we blush, we can gain a valuable window into the complex and puzzling relationships between our mind, body and society.

Blushing occurs when the small blood vessels that supply the skin widen thus allowing an increase in blood flow. 'Blushers' report a burning sensation in their face and often a full body tingle. In most cases, blushing passes within a few seconds to five minutes. According to Dr Roger Dampney, Reader in Physiology at Sydney University, blushing is a widespread phenomenon which science knows 'surprisingly little about'. Furthermore, 'it is triggered by emotional stimuli', 'higher levels of the brain are involved', and there is some anecdotal evidence which suggests that blushing may not be confined to the face.

Blushing is one of only a very few body changes triggered directly by the mind. It appears to be biologically driven rather than learned, but definitely socially induced. People do not blush in private. And you can make someone blush merely by accusing them of already blushing. Another curious thing about blushing is that people can blush and be made to feel embarrassed even when they have done nothing wrong. Simply being conspicuously different, good or bad can raise a blush. Being complimented or overly praised is an example.

Interestingly, even people who have been blind since birth are known to blush. And if you tell a person they are starting to blush, chances are they will. In fact, researchers use this technique when attempting to induce blushing in order to study it. But at the same time, it is virtually impossible to make yourself blush. Chronic blushers sometimes seek therapy to overcome the problem. They are often coached that when they feel a blush coming on, they should try to make themselves turn as red as possible. It has been reported that this stops the blushing completely in many cases.

Despite theories going back to Darwin and Freud, no-one is absolutely certain why it is only humans who blush. Nevertheless, one simple explanation continually pops up: humans are the only primates with their facial skin completely exposed. Although other primates may also blush, it is visible only in humans. Another simple explanation is that blushing requires a person to have a sense of self, embarrassment capability, and the ability to judge themselves from the viewpoint of others. Since no other animal can do this to the degree that humans can, blushing is uniquely human.

Charles Darwin (1809–1882) devoted an entire chapter

to blushing in his book *The Expression of Emotions in Man and Animals* (1872). Darwin was the first to note that blushing is a trait exclusively human, universal among all peoples of the world, and also characteristic of the blind. He argued that blushing was almost certainly an inherited trait and was triggered by the attention of others. Darwin wrote, 'It is not the simple act of reflecting on our own appearance, but the thinking what others think of us which excites the blush.'

In *Inhibitions, Symptoms and Anxiety* (1926), Sigmund Freud (1856–1939) argued that blushing is a complex reaction which results from repressed sexual excitement, exhibitionistic wishes, and surfaces in the face of the fear of castration. It is an indirect way of conveying to others one's erotic urges. To Freud, a blush typified the internal subconscious power struggle between the id and the superego.

Even today, some Freudian psychologists still adhere to this basic point of view. For example, Dr Carole Lieberman, a psychiatrist in Beverly Hills, California, puts it this way: 'A gratified sexual wish is embarrassing. When a woman crosses a vent at an amusement park, her skirt flies up. She blushes. She may blush not simply because she is exposed but also because she had an inner wish to exhibit herself.' Dr Sydney Feldman, a New York psychiatrist, puts it even more simply: 'Men blush because they feel castrated, and women blush because they are not men.'

Blushing is in many ways strangely contradictory. According to Dr Murray Blimes, a psychologist in Berkeley, California, 'the striking thing about blushing is its implicit mixed signals. A blush is a funny mixture of wanting to hide and at the same time wanting to attract someone.' He notes, for example, that in a group of three, when one person discloses to a second person that a third person has a secret passion, the third person blushes. As Dr Blimes says, 'something has been exposed that they want to hide. Yet the reaction is partly to hide, but also to confirm it. There is a funny dialectic going on. That is a very important part of a blush.' He adds, 'blushing is literally waving a red flag at a bull, a come on, a variation on the fight-or-flee response. Blushers want to hide, yet the blush draws attention to themselves.'

Nevertheless, a new theory for why we blush has recently emerged. It holds that blushing is actually an instinctive way to get back into the good graces of others. It is an attempt to avoid being ostracised from a group for breaching unwritten rules of society. The author of what is now known as the 'appeasement theory' of blushing is Dr Mark Leary, a professor of social psychology at Wake Forest University in North Carolina. Dr Leary first presented his theory at a 1990 meeting of the American Psychological Association.[22]

He argues that we blush 'when we have done

something that threatens our status in a group, when we have done something deviant. Blushing is an appeasement behaviour to defuse a potentially ugly situation. Moreover, most animals have ways of doing this. For example, when other primates are threatened by a dominant animal they lower their eyes and get this really cheesy grin on their faces. Sometimes they avert their eyes, take on sheepish grins, or present their rear ends. It defuses the likelihood that they will lose status in the group or suffer aggression.' Dr Leary says he notices that same 'cheesy grin' on the faces of people who are blushing. While other primates use those instinctive gestures to allay aggression or avoid banishment, Dr Leary claims that humans use them for the latter. He observes, 'blushing is saying to everyone else, "Oops, I recognise that I've broken a social rule". It is like a non-verbal apology – an instinctive acknowledgement that one has done something wrong – its purpose is to re-endear a person to a group in the face of impending exclusion.'

Regardless of whether Dr Leary is correct, there is some evidence that blushing works as Dr Leary contends. At least blushing seems to create more empathetic feelings toward the blusher.

According to Dr Rowland Miller, a psychologist at Sam Houston State University in Huntsville, Texas, several studies suggest the power of the blush. For example, in one British study he cites, viewers watched a videotape of a clumsy shopper accidentally knocking over a supermarket toilet paper display. The researchers offered three alternative endings, asking the viewers what they thought about the culprit after each. In the first ending, the shopper merely flees. In the second, the seemingly unembarrassed shopper calmly rebuilds the display. In the third, the obviously mortified shopper blushes, looks around sheepishly, and eventually gathers up the wreckage. Dr Miller notes that the viewers expressed the greatest degree of warmth towards the blushing shopper.[23]

Thus, Dr Miller concludes that a blush – the universal signal of human embarrassment – produces empathy while lessening hostility.

Dr Blimes observes that blushing 'is really very interesting. Of all the possible things that separate animals from humans, the blush is our most distinctive characteristic. Why in the world did nature elaborate this thing?'[24]

¶ Why do I have fingerprints?

This OBQ is one of the more popular. 'Vestigial' is the medical term given to a feature of the body that has no purpose. Fingerprints are certainly not vestigial.

Fingerprints are visible parts of the rete ridges where the skin's epidermis dips down into the dermis forming an interlocking structure. Our unique fingerprint and toeprint configurations are due to the semi-randomness of ridge and dermal structure growth.

Fingerprints help us to grip and handle objects in a variety of conditions. They work on the same principle as automobile tyres. We have evolved a system of troughs, ridges, and grooves to help channel water away from our fingers and toes. Although it doesn't seem like much, this results in a somewhat better gripping ability. Toeprints do the same thing. They help us keep our balance and prevent us from water-planing over smooth, wet surfaces.

Fingerprints also protect us against blisters. Fingerprints help alleviate the sideways stress that would otherwise separate the two layers of skin and allow fluid to accumulate in the space and thus form a blister.[25]

¶ What turned Michael Jackson's skin white?

Dudley Moore had it, and Steve Martin may too. Michael Jackson says that treatment for this strange disease has bleached his skin white. It is called vitiligo and much mystery surrounds it.

Vitiligo is a disorder in which the skin loses pigment due to the destruction of pigment cells called melanocytes. Areas of the skin become white. The pigment loss is not over the entire body, merely in patches. The most common patches are around body openings (such as the eyes), body folds (such as the armpits or groin), or exposed areas (such as the face and hands).

Vitiligo can strike either gender and at any age but it usually shows itself before the age of 20. Vitiligo is quite common as well. Perhaps 1–2 per cent of the general population has it to some degree, although it may be confused with other skin problems. It is more common in people with thyroid conditions and certain other metabolic diseases. It is also more apparent in darker skinned people. Yet most people who have vitiligo are in otherwise good health and suffer no symptoms other than the blotched areas of pigment loss.

Vitiligo is not contagious, nor is it in any way associated with leprosy. The old description of vitiligo as 'white leprosy' has no basis in scientific fact.

Once vitiligo patches appear, there is no way to tell if

they will increase in size or number. In many cases, initial pigment loss will occur but stabilise thereafter. In others, pigment loss can fluctuate. Psychological factors may play a part in such fluctuations as many patients report that they had experienced initial or subsequent episodes following periods of physical or emotional stress. A theory given serious consideration is that stress somehow triggers the depigmentation process in the human cell among those who are already genetically predisposed. Interestingly, sometimes depigmentation patches may spontaneously repigment. Why this happens remains a mystery.

Medical researchers are not sure what causes vitiligo. Some assert that the body may develop an allergy to its own pigment cells. Others argue that the cells may strangely destroy themselves during the process of pigment production.

A common fear among vitiligo sufferers is that it is linked to skin cancer – that it is an early warning sign. However, there is no causal link between the depigmented patches and either cancerous or pre-cancerous conditions. But skin cancer patients sometimes develop vitiligo after their skin cancer symptoms appear. The reasons for this are unclear. And even more odd, in many skin cancer patients who develop vitiligo, the vitiligo seems to actually stop the spread of cancer. Why this happens is also rather baffling.

Vitiligo can strike anyone. In somewhat more than half of all cases there is a family history of the disease. Often too, there is a family history of early greying of the hair as well. Statistically, early greying seems to foreshadow vitiligo and vice versa. Sometimes a vitiligo sufferer will not realise that there is a family history of the disease. Instead, they think that there is merely a family history of early greying.

The good news is that vitiligo can be treated. In mild cases, make-up will cover the blotches without any treatment. Moderate cases respond to ultraviolet light, steroids, and especially to drugs such as psoralen – alone or in combination. Such therapy aims to repigment white patches to a darker colour. Repigmentation seems to work best when only a few small white patches exist.

Presumably, in Jackson's case the repigmentation failed. When it fails in the most serious vitiligo cases, then depigmentation is attempted. Monobenzone is used to bleach the body's entire skin surface so as to give the patient at least an even, all-over colour. Under a physician's careful instructions, monobenzone is applied two or three times daily until the bleaching is complete and then twice weekly afterwards.

Jackson is almost certainly using monobenzone because it is 'the only treatment' to depigment. This is according to Dr James Nordlund of the Department of Dermatology at the University of Cincinnati School of

Medicine. Dr Nordlund adds that monobenzone is to be prescribed with care and only in 'patients with very extensive vitiligo'. Furthermore, the drug 'occasionally' produces irritation, takes about six to 12 months to fully take effect, and has about a 75 per cent success rate.[26]

If Jackson stops his monobenzone treatments, his previous colour will return. In fact, Michael Jackson could turn back to black whenever he wants.[27]

7 · **The Hair and Nails**

In The Rape of the Lock, *poet Alexander Pope (1688–1744) wrote the immortal lines, 'Fair tresses man's imperial race ensnare, And beauty draws us with a single hair.' If ol' Alex didn't have most of us in mind with that one, he was awfully spot-on just the same. Hair has been an important part of our human life and culture. The Samson story from the Bible, the Rapunzel fairy tale and on and on...*

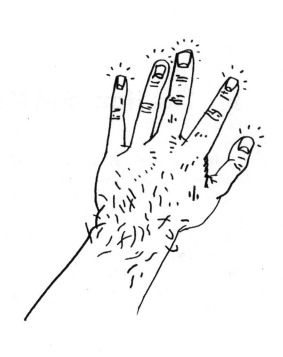

¶ Why does my hair turn grey?

This is one of the more often asked OBQS. In the vast majority of cases, age and the erosion of melanocyte functioning cause hair greying.

The greying process starts deep in the outer layer of the skin (epidermis) all the way to the inner layer of the skin (dermis). Each of the 100,000 hairs or so on the average human head is controlled by a hair bulb located below the hair follicle. It is through the hair bulb that a variety of complex chemical substances are channelled which create each hair. These substances are mainly composed of keratin.

Millions of melanocytes (protein-producing pigment cells) located at the hair roots and in the epidermis produce chemicals that determine the colouring of both hair and skin.

Those who suffer from albinism usually have a normal number of melanocytes but, due to a genetically caused deficiency, they lack the chemical means to trigger pigment production. In some people, only a small area of skin lacks functionless melanocytes. This produces white spots or streaks in an otherwise dark head of hair.

The melanocytes yield the colour by 'treating' the hair at the follicle. Eventually, the hair grows long enough to reveal the colour. Once the hair has been pigmented by the melanocyte's action, the colour cannot be altered.

This is because the pigmentation is not just a coating to the keratin body of the hair, but is infused. The pigmentation chemical, melanin, has two basic components. These components predispose a hair to be dark, light, or some hue in between, depending on the proportion of each pigment. Greying occurs when there is an age-related shutdown of the melanocytes. This usually occurs gradually over several years. Although greying can occur quickly, you cannot turn grey overnight. Your hair does not fall out that fast. Some people report that they turned grey overnight. However, closer examination of such cases invariably shows that they were turning grey slowly but just didn't notice the change. Their awareness of being grey is what happened overnight.

About 100 hairs a day are lost from natural attrition. With age, the older dark hairs fall out, leaving a greater proportion of newly-created white hairs. As white hairs gain the majority, the greyness appears to increase. Thus, greyness is an optical illusion created by the mixture of the remaining dark hairs and the newer white hairs.

Greying is almost certainly largely genetically determined. Nevertheless, stress and worry can probably influence the rate at which a person turns grey. Men and women grey in slightly different patterns.

Women grey slightly faster than men. Premature greying affects about 25 per cent of all people by the time they are 25 years old. Production of the first grey hairs begins at about age 15. Ironically, the colourising cells often speed up pigment production as we age, so hair sometimes darkens temporarily just before the pigment cells die.[1]

Thyroid disorders are a common cause of premature failure of melanocyte functioning.[2] Diseases affecting the pituitary gland probably reduce hair colouration, as can interruptions of hormone production in the testicles or ovaries. Diabetes can upset melanocyte functioning as can severe malnutrition. Premature greying has also been associated with a possible increased risk of heart disease. Some evidence suggests that in cases of pernicious anaemia, a deficiency of vitamin B12 may influence melanocyte functioning and thereby influence greying.[3]

¶ Why is the hair of some people curly and others' straight?

The pattern of hair can be naturally straight, curly or something in between. Our genes ordain not only our hair colour, but also our hair pattern. Small communities that are endogamous (marry inside the group) will, over many generations, evolve a single hair pattern. An example of this is the San people (sometimes called the Kung Bushmen) of the Kalahari Desert of Southern Africa. All San have 'peppercorn' hair. As this vivid term implies, their hair is very, very tightly curled.

¶ How does a perm make my hair curly?

The chemicals in a perm make hair curly by first breaking
down and then reforming the actual chemical bonds
of hair.

Hair is composed mostly of a protein called keratin.
Keratin contains large amounts of sulphur. Chemical
bridges connect sulphur atoms from one protein
molecule to nearby sulphur atoms in neighbouring
molecules. Like rivets holding together the steel girders
of a skyscraper, these chemical bridges hold the keratin
molecules together in a structure that gives each hair its
characteristic shape.

One can bring about a temporary alteration of the
hair's natural pattern in several ways and for varying
lengths of time. For example, by tightly wrapping straight
hair around curling irons, one can hold the curl for a short
time. However, the alignment of the hair's proteins will
eventually force it back into its naturally straight shape
pattern.

On the other hand, when you get a perm, the proteins
in the hair are allowed to realign so that they favour a
curly shape instead of the straight shape. The hairdresser
first winds the hair around the curlers, and then applies
a chemical that breaks down the bridges between the hair
proteins. Just as steel girders can slide when their rivets
are removed, the proteins can now slide with respect to

each other. Freed from their former restraints, the proteins settle into positions that conform to whatever shape the hairdresser determines. If the hair is wrapped around curlers, curly hair is what results.

Nevertheless, there is one final step. The re-ordering must be 'locked in'. The hairdresser eventually washes away all of the bond-breaking chemicals and adds a different chemical solution that causes new bridges to form. This fixes the proteins in their new, curly configuration. Even wind, sun, and shampooing cannot alter this rearrangement. It is only after our hair grows out again that our hair resumes its original genetically-determined pattern.

As powerful a solution as a perm is, it cannot alter our genes.[4]

¶ Why do only men have beards?

This is a matter of hormones. Facial hair is linked to testosterone, the male hormone. Genes determine our level of testosterone. Males have more testosterone than females, hence usually more facial hair. Studies have shown that bearded men are 'rated significantly higher on masculinity, aggressiveness, dominance, and strength'.[5]

The ancient Iraqis were the first true hair stylists. Then known as the Assyrians, inhabiting what is now northern Iraq, the skill of Assyrians at cutting, curling, layering, and dyeing hair was legendary throughout the Middle East. Vidal Sassoon would have every right to be green with envy – except that this was some 3,500 years ago. Nevertheless, this ancient Assyrian craft grew out of a love for hair – an obsession with hair – especially for soldiers in battle.

The Assyrians cut hair in graduated tiers, so that the head of a fashionable general was as neatly geometric as an Egyptian pyramid – and somewhat similar in shape, although upside down. Longer hair was elaborately arranged in cascading curls and ringlets, tumbling over the shoulders and onto the chest.

Hair was oiled, perfumed and tinted. Men cultivated a neatly clipped beard, beginning at the jaw and layered in ruffles down so far as to completely cover the neck. Kings, generals, and lieutenants had their abundant, flowing hair curled by slaves, using a fire-heated iron bar – the first curling iron known to history. No self-respecting warrior would dare enter battle without seeing to it that his hair was properly done. Hair, like armour, had to be worn properly and with distinction. If he fell, he fell in style.

Indeed, so obsessed with hair were the Assyrians, that they developed hair styling to the exclusion of nearly every other cosmetic art. Law even dictated certain types of coiffures according to a person's rank – in military and civilian life.

One rule prevailed: baldness, full or partial, was a sign of weakness, impotence – the mark of the eunuch. Baldness was considered by the Assyrians as an unsightly defect to be concealed by wigs in public.

Hair styling was extremely important to Assyrian women as well. As was the case in ancient Egypt, high-ranking women, during official court business, would appear only after donning stylised fake beards to assert that they too could be as authoritative as men – the beard was the symbol.

It was only later, with the ancient Romans, that beards and long hair fell out of favour as appropriate for the battlefield. Prior to battle, the Romans shaved with crude razors. It was thought that in hand-to-hand combat a long beard and long hair could be used to pull a man down.

In fact, the Romans used the word for 'beard' (barba) as the term for their enemies in battle. Hence, we have today the word 'barbarian' as well as 'barber'.

¶ Why do some of us pull out our hair?

Many of us pull or twirl our hair. Sometimes this is a habit we develop in childhood. But some people compulsively pull their hair out, often including their eyebrows, eyelashes, underarm hair, even pubic hair. It is a bizarre condition called trichotillomania.

Mental health professionals usually classify trichotillomania as an 'impulse control disorder'. Although chronic hair pulling will not kill anyone, some individuals cause themselves injuries and pain, and risk disfigurement if they persist in giving in to this impulse. Moreover, the condition is associated with high rates of what is termed psychiatric 'co-morbidity'. That is, other conditions happen along with it. For example, trichotillomania is believed to lead to marked emotional distress and diminished self-esteem.

However, just because someone often pulls their hair, it may not be trichotillomania. The internationally accepted standard of the *Diagnostic and Statistical Manual of Mental Disorders* (DSM-III-R) published by the American Psychiatric Association in 1989 recommends that doctors should look for five signs or clinical criteria before being sure it is really trichotillomania:

|| The patient must recurrently fail to resist impulses to pull out their hair, resulting in noticeable hair loss.

|| The patient must experience an increasing sense of tension immediately before pulling out the hair.

|| The patient must feel gratification or a sense of relief when pulling out the hair or immediately afterwards.

|| The patient must have no pre-existing inflammation of the skin.

|| The patient's hair pulling must not be a response to any delusion or hallucination.

Nevertheless, although most of us probably know a 'hair puller' – it is often seen in children – for the last thirty years, trichotillomania has generally been considered to be a rare condition. For example, Dr S. Muller of the Mayo Clinic in Minneapolis reported that there are only 15 cases per year among the thousands of patients seen for various complaints at the world famous Mayo Clinic. Moreover, it is widely believed that trichotillomania is more common in women than in men – at a ratio as high as nine to one. Indeed, Dr Muller found that three out of four trichotillomania sufferers are women.[6]

But in 1991 three psychiatrists from the University of Minnesota claimed that conventional wisdom was wrong about trichotillomania. Drs Gary Christenson, Richard Pyle and James Mitchell found that 'trichotillomania may not be as rare as previously suspected and may affect males almost as often as females'.[7]

They base this conclusion on their study of 2,579 university students. They found that 1.5 per cent of men and 3.4 per cent of women had visible hair loss. This is despite the fact that less than half of these chronic hair pullers met all five trichotillomania criteria. But interestingly, when all the criteria were met just as many men as women qualified as true trichotillomania sufferers. Furthermore, the researchers stress that theirs was not a clinical population of patients but a more general group of people, which is all the more surprising.

If these doctors are right, then trichotillomania probably occurs more often in the general population than it does among people treated for mental illnesses. And even though men have shorter hair – the common reason given for men exhibiting less trichotillomania – they may suffer from it nearly as often as women.

Why do people engage in chronic hair pulling? Of course we don't know. It is one of the mysteries of human behaviour if not the body.

Dr Judith Rapoport of the National Institute of Mental Health in Washington suggests some possible hints. Perhaps trichotillomania has important symbolic value, hence is the chosen form for a compulsion.

In classic psychoanalytic literature, hair has multiple symbolic meanings. It is a symbol of beauty, femininity, virility and physical prowess. Dr Rapoport writes, 'It is also a bisexual symbol and sexual conflicts are said to be displaced to the hair.' Haircutting, pulling, or plucking is felt to symbolise castration or the rendering of oneself as infertile or impotent.

For the victims of the Nazi concentration camps, the shaving of the head was considered to be one of the most excruciating of tortures and debasements.

Moreover, Dr Rapoport relates that a psychoanalytic essay by Dr Edith Buxbaum on the fairytale Rapunzel suggested that the cutting of Rapunzel's hair represents separation and loss of the mother. Furthermore, in various Hindu cultures shaving the scalp is associated with mourning. In some areas of India, hair is plucked before a person enters a life of penance, and there is a long tradition of shaving one's head in the Christian monastic life.

Although we have so far only poor explanations for what causes trichotillomania, what is known is that those who suffer from it are rendered utterly miserable.

Fortunately, the news is far better about the treatment of trichotillomania. Talk therapy often achieves a marked improvement. Support groups exist and are believed to offer great help. Drug treatments such as the use of Clomipramine have often led to dramatic results. In one study, nine out of 10 trichotillomania sufferers got 'much better' through the use of this very potent antidepressant.[8]

Drug therapies are continually being improved. Ironically, it is hope for those with trichotillomania that is 'at hand'.

They need no longer tear out their hair – in more ways than one.

¶ If I lose my hair can it ever grow back?

While a few of us pull our hair out, many of us want to put it back. A new experimental product may help us retain our youthful hair long after it would normally lose its body and sheen. Losing hair as a result of age is not just a male problem. About 30 per cent of women also suffer from age-related hair loss by the time they reach 40 years of age. It is called 'pattern alopecia'. However, the new product is already a popular prescription drug but, oddly enough, a diruetic. It is called spironolactone.

Research at the Philip Kingsley Trichological Centre in London claims to have shown that spironolactone 'greatly improves' pattern alopecia. A Kingsley endocrinologist gave oral doses of spironolactone to six women between the ages of 30 and 45 who were suffering from moderate to severe hair loss. In four of the women the drug stopped further loss, while in two others it unexpectedly promoted hair growth. The only side effect of spironolactone seems to be a slight menstrual irregularity.

Spironolactone is known to reduce the activity of testosterone and other male-associated hormones. And it is known that female pattern alopecia is triggered by a sensitivity to these hormones.

The researchers recently expanded the trial to include 25 more women and 12 balding men. In the near future, they hope to test spironolactone on hundreds of human subjects if their results continue to be so promising.

Dr David Kingsley, the founder and director of the centre, claims that, 'We haven't found the magic oil that will grow full heads of hair, but clearly spironolactone is well worth investigating further.'[9]

¶ Why do I bite my nails?

The medical term for nails is 'unguis'. It is unclear why so many of us bite our nails. Some people even bite their toenails. The most common theory is that nail biting is stress related. This notion holds that nail biters alleviate built-up, everyday tension through the unconscious and unintentional biting or peeling of nails. Thus, nail biting functions similarly to scratching the head, pulling the hair, cracking the knuckles, and so on. Another theory, a neo-Freudian one heard less often these days, is that nail biting is a substitute for masturbation. Hence, nail biting emerges since it is both less heavily repressed and socially tabooed than masturbation. Still another view holds that nail biting reflects nutritional deficiencies, especially where the nail shaving is swallowed. Nails are made of a particularly strong protein called keratin. Hair is made of this same protein.

The medical term for nail biting is 'onychophagia'. But far from being merely unsightly, onychophagia can have health consequences too. According to Drs Peter Samman and David Fenton, onychophagia can make one vulnerable to skin infections. Moreover, in severe onychophagia, when the nails are bitten all the way down to the whitish, semicircular 'half moon' region near the cuticle, permanent nail deformity can occur.[10]

Nail biters normally try numerous remedies to curb

the biting or peeling urge. These include bandages, aversion therapy using hot sauce or bitter-tasting nail coatings, relaxation tapes, hypnosis, chewing gum, and many others. 'Sadly, these remedies seldom work', according to Dr Richard Scher, professor of dermatology and director of the Nail Research Center at Columbia University in New York. Dr Scher, along with Dr Dale Daniel, is the author of perhaps the most influential work on this topic.[11]

Dr Scher notes that nail biters usually chew their nails unconsciously. Thus, they cannot stop until they develop a keen awareness of when, and under what circumstances, they are most prone to bite. Indeed, 'they should cultivate an awareness of where their hands are at all times'.

It is self-defeating when parents punish children who bite their nails. This tends only to increase tension and make matters worse – perhaps resulting in even more nail biting. Instead, parents should point out to children when they are biting their nails and in a supportive manner encourage them to be more aware of their hands. It also helps to reduce stress in the child's life.

Dr Scher claims that several myths exist about nails and nail biting:

|| *Manicures make no difference.* 'To the contrary, nail polish makes nails more attractive, and the

better-looking our nails, the less likely we are to abuse them.'

|| *Nail polish is for women only.* 'In fact, a growing number of men are now having their nails manicured. And clear polish is virtually undetectable.'

|| *Gloves make no difference.* 'Covered nails are not only less tempting than uncovered nails, but also are harder to reach if an urge to bite proves irresistible.' Furthermore, 'for maximum protection, wear gloves at all times – day and night – until the nail biting urge dissipates. If you have the urge and are not wearing gloves, lightly clench your fists until it passes.'

|| *Psychotherapy does not work.* 'In particularly severe cases of nail-biting, short-term therapy may be helpful. In some cases, tranquillisers may be prescribed.'

The practice of polishing, painting, or otherwise adorning nails goes back to the ancient Chinese, 5,000 years ago. In ancient China, both women and men used a paint made of beeswax, egg whites, gelatin, and gum arabic. At roughly the same time, the ancient Egyptians adorned their nails as well. Egyptian women and men of high social standing used henna to stain their nails a rich, red-orange – the deeper the red, the more exalted and important the individual.

Like hair, nails are composed of dead tissue. That is why it does not hurt when hair or nails are cut – only if pulled. This is because the living growth is deep within the skin.

Fingernails grow at the speed of about 4 cm per year. Fingernails grow two to three times faster than toenails. Few factors can speed the rate of nail growth. Nevertheless, it is known that nails grow faster during hot weather compared to cold, grow faster during the day than at night, and grow thicker as we age. Hot weather slightly speeds the rate of growth since heat increases the rate of all metabolic processes.

Generally, nails of males grow slightly faster than nails of females. Interestingly, women can expect a slight spurt in their normal nail growth rate just before menstruation and throughout pregnancy – a response to hormone activity.

It is a myth that nails grow after death. It merely looks that way since skin dries, shrinks and retracts from the ends of fingers and toes.

Oh yes, death is a proven cure for nail-biting – although somewhat extreme.[12]

¶ Why are some people so hairy?

Excessive hair occurs in 10 per cent of adolescent girls and an equal number of women during the child-bearing years. The hair grows on the face, chest, back, buttocks, linea alba (the line between the navel and the pubic region) – and elsewhere.

These are places where current fashion standards claim only males are supposed to have hair. So when even a little hair appears on a female, some girls wonder if they are 'turning into a man'. They are probably confusing the more common condition of hirsutism with the far less common condition of virilisation.

Hirsutism is the medical condition where there is an excessive amount of body hair. Of course, just what constitutes 'excessive' is highly subjective. There is no standardised medical definition.

Dr Gordon Senator, an endocrinologist at the Royal Hobart Hospital draws upon one of the few studies of hirsutism rates. In a 1964 UK report in *The Lancet* by Dr E. McKnight of the University of Wales, a survey of 400 women students found that 26 per cent had 'terminal hair' on the face. In these individuals, more hair appeared on the upper lip than on the chin. Furthermore, facial hair was clearly noticeable to an 'impartial observer' in 10 per cent of the women. Finally, in four per cent of the women, hair was sufficient enough to require treatment in the opinion of Dr McKnight.

Hirsutism differs from virilisation. Virilisation is the induction of increased body hair, but also deepening of the voice, increase in muscle mass, development of a hairline typical of the male forehead, stimulation of secretions and proliferation of the sebaceous glands often leading to acne, hypertrophy of the clitoris (the clitoris becomes enlarged), oligomenorrhoea (menstrual periods become erratic and flow is reduced), and/or amenorrhoea (periods cease).

Both hirsutism and virilisation are caused by an overabundance of testosterone (male hormone). Usually this is genetically based, but sometimes diseases or other medical conditions may be involved. Polycystic ovary disease is perhaps the most common of these.

In the classic text, *Hirsutism* (1981) by Drs P. Mauvais-Jarvis, F. Kuttenn, and I. Mowszowicz, three endo-crinologists point out that 'there is a direct relationship between the extent of hirsutism and the level of plasma testosterone in hirsute women'. In general, the more male hormone, the greater the degree of virilisation. However, the three doctors observe that, interestingly, 'in most cases of hirsutism, plasma testosterone is either only slightly increased or within the normal female range'.[13]

Hirsutism does not indicate a so-called lack of femininity. 'Femininity', based upon the current standard of feminine beauty dictated by the fashion industry, is highly subjective. What is considered beautiful in one historical period may not be in another. In fact, during Victorian times, at least a moderate degree of hirsutism in a lady was considered quite fashionable and very attractive. Men's views about women's body hair are quite diverse. This is similar to women's varying preferences for facial or chest hair on men.

The degree of hair growth varies among ethnic groups. Northern European women tend to have less hair than Southern Europeans. Asian women have less still. Moreover, birth control pills can cause extra hair growth by altering normal hormone levels.

Should hirsutism be severe, it may be treated. According to Drs Stephen Judd of the Flinders Medical Centre in Adelaide and John Carter of the Concord Hospital in Sydney, cosmetic treatments include 'bleaching, plucking, depilatory creams, waxing, shaving, and electrolysis'.[14] Various hormone therapies can also be used. Hormonal treatment of hirsutism should not affect libido or other aspects of sexuality.[15]

Hirsutism does not become worse if a woman removes the hair by shaving. In fact, hair is created far below the skin's surface. Shaving does not affect the rate of hair growth nor its coarseness.

At times a young woman may turn to a man, but never into a man.[16]

¶ Do humans get hair balls like cats?

Not exactly, but your own hair can kill you. Hair bezoars (trichobezoars) are pathological masses of swallowed hair that form, harden, and become lodged in the human gastrointestinal tract. We share the capacity to form bezoars along with such animals as goats and Persian cats. Human bezoars can cause abdominal obstruction, haemorrhage, and perforations. Death results in 30 per cent of diagnosed cases left untreated. A bezoar-related death is often prolonged and excruciatingly painful.

Hair bezoars occur occasionally in people who chew their hair, most frequently in young girls and women. Inevitably, some hair is swallowed. If this is done habitually, a hair bezoar can result. The person suspects nothing until symptoms emerge. Hair bezoars are less of a public health problem these days since hair styles for both males and females are shorter.

Medical treatment for human bezoars can include endoscopy (viewing the bezoar through an inserted small, flexible fibreoptic tube), injecting the bezoar with an enzyme solution to fragment it or surgery.

There are two other types of human bezoars: phytobezoars and trichophytobezoars.

A phytobezoar is composed of vegetable matter such as citrus-fruit pulp, fibres, skins, and seeds. Again, too much of this material is swallowed. Of all fruits, the skin of the persimmon seems to be the most prominent in forming phytobezoars. A trichophytobezoar is a combination of hair and vegetable matter, hence its name.

Articles on bezoars surface in the medical literature from time to time. One of the more memorable of these was written by three Belgian doctors.[17] Colourfully terming it 'The Rapunzel Syndrome', the doctors describe an unusual case of the trichobezoar in a fourteen-year-old girl who loved to eat her stuffed, soft-toy animals as well as the family's carpet.

If it seems bizarre that humans would swallow such things, remember that there are at least 31 cases in the medical literature of humans swallowing just one everyday household object – the toothbrush.[18]

But the definitive article on human bezoars may be found in the *Medical Journal of Australia*.[19] Dr Randolph Williams, visiting surgeon at the Royal Adelaide Hospital, writes that bezoars (from the Arabic badzehr meaning 'poison antidote') were once highly prized throughout Europe. They were thought to contain great medicinal properties. Indeed, the most valued bezoar of all was that obtained from the fourth stomach of the Syrian goat. Down-market, second-rate bezoars from other animals and humans were often substituted as counterfeit ring-ins. This was because Syrian goat bezoars were always in short supply. During Medieval times, a trade in bogus bezoars flourished. Many tests were devised by the wisest of men to detect the fakes.

Bezoar stones were carried as charms to ward off illness and poisoning. For example, at the king's table guests would be served wine in a jewelled goblet. Attached to the goblet, hanging from a small chain would be an equally bejewelled bezoar. Thus, it would be handy for dipping (similar to a tea bag of today). In addition, bezoars were often pulverised to a fine powder and administered internally as a treatment for many medieval maladies.

A bezoar stone was set in gold and included in an inventory of crown jewels belonging to Queen Elizabeth I. As Dr Williams writes, the Queen's high esteem for bezoars 'attested to its popularity with English royalty'.

Bezoars remained in medical use until the mid-eighteenth century when they were finally superseded by more effective treatments.

Despite the high cost of medical care these days and the movement towards traditional folk medicines, it is unlikely that bezoars will make a comeback as a standard form of therapy.

But if you notice a Syrian goat grazing at the back of your doctor's surgery...[20]

¶ If my hair is dead, why does my physical and emotional state change its condition?

This OBQ comes up frequently along with queries about how hair conditioner works.

Although dead, the appearance of hair depends on secretions from the sebaceous gland at the base of the hair follicle. Too little of the secretions and the hair becomes dry and brittle, and looks damaged. Too much of the secretions and it becomes greasy and lifeless (this sounds like a shampoo commercial!). Sebaceous secretions are affected by age, general health, hormone flow, and even emotional state. Pregnant women often notice changes in their hair. This is due to the hormone changes associated with pregnancy.[21]

8 · The Skeleton, Bones and Teeth

The human skeleton is a wonder of engineering. You could compare it to a strong building. For example, kilo for kilo or pound for pound, your femur (thighbone) is stronger than reinforced concrete. When you walk fast, your femur resists an average of 80 to 90 kg per square centimetre. Most of the world lives in houses less sturdy than this! Bones consist of calcium and phosphorus so densely packed that the crystalline pattern of bone resembles that of diamonds – the hardest natural substance known. And let's not forget the strength of teeth. Tooth enamel is the hardest substance produced by the human body. It is so hard that a dentist's drill has to turn at 1,000,000 rpm just to cut through it.

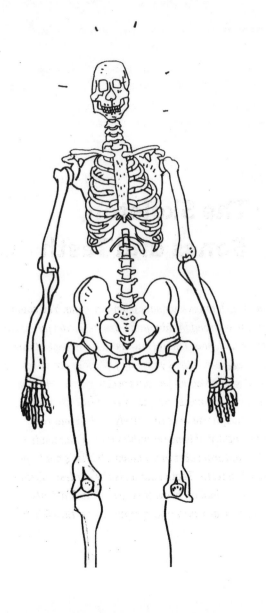

¶ How many bones are in the human body?

You're bound to be asked this by somebody sometime or another. You're right if you answer 208. The answer is not likely to change in the next few years, so you can go ahead and memorise it. Where are the various bones located? Well, more than half are located just in the arms and legs, if you count the hands and feet with them.

The bone breakdown goes like this: 60 in the upper extremities (arms and hands), 60 in the lower extremities (legs and feet), 26 in the vertebrae (backbone), 24 vertebral ribs, 22 in the skull, six in the ears, four in the pectoral girdle (collar and shoulders), three in the sternum (breastbone), two in the pelvic girdle (hips), and one in the throat.

You're also bound to be asked by the same person on the same occasion: how many teeth are in the human body? Go to the head of the class since you just correctly answered 32.

If you already know so much, why are you reading this book?

¶ How heavy can a person be?

Although scientists debate this, there is probably no limitation on how large humans can become. But already, we are capable of doing things in big ways – very big ways.

The world's heaviest human ever to have lived was Jon Brower Minnoch (1941–1983). He weighed more than 629 kg at his heaviest.

In March, 1978, the 185.4 cm Minnoch checked into the University Hospital in Seattle. He was saturated with fluid and suffering from heart and respiratory failure. Twelve fireman were required to move him and two hospital beds lashed together were needed to hold him. Dr Robert Schwartz, a consultant endocrinologist, calculated Minnoch's weight based upon his intake and elimination rates. He recovered and reduced his weight to a low of 215.8 kg, but much of the weight returned. When he died in 1983, Minnoch weighed more than 361.9 kg.

The world's heaviest surviving human baby by vaginal delivery was a boy weighing 10.2 kg. He was born in 1954 to Signora Fedele of Aversa, Italy. The heaviest baby by caesarean section delivery was also a boy who weighed the same. He was born in 1982 to Christina Samane of Sipetu in the Transkei, South Africa.

Larger babies have been stillborn. The largest weighed 13.26 kg, and was born in Effingham, Illinois in 1939.

A baby is not the heaviest thing a human body can produce. A cyst can be far heavier. On 24 October 1991, a gigantic 137.4 kg cyst was removed from the ovaries of a thirty-six-year-old Californian woman. When the massive abdominal growth was removed, it had to be carried out on its own stretcher. The woman is now fully recovered and leading a normal life.

Dr Katherine O'Hanlan, a specialist in gynaecological oncology, lead the surgical team at the Stanford Medical Center in Palo Alto, California. Dr O'Hanlan said that the ovarian growth was by far the largest cyst in history. In fact, it outweighs the next largest by more than 53 kg.

The patient is 178 cm tall and weighed 95 kg immediately following the operation. She had been bedridden for two years prior to surgery due to the great weight of the non-cancerous cyst.

Dr O'Hanlan said the cyst, which had grown to more than 92 cm in diameter over a ten-year period, was 'highly unusual'. Most ovarian cysts grow to no more than 20 cm, she said.

Dr O'Hanlan added that 'the patient had read about the difficulties of removing cysts much smaller than hers, and was wisely hesitant to have surgery. She was afraid of losing her life. But things had just gotten to be unmanageable. She was caught between a rock and a hard place.'

The delicate operation took more than six hours. The

mass protruding from the woman's lower abdomen was made up of many small cystic structures fused together 'like a water balloon that contains many other balloons', Dr O'Hanlan said.

It 'had to be carefully trimmed away from the abdominal wall and the bowels. We rolled it onto a stretcher so it could be sectioned or cut up to be examined under the microscope. None of us could lift it,' Dr O'Hanlan added.

Previous to this, the largest cyst ever removed from a patient was an 83.9 kg growth removed from a patient in the US state of Maryland earlier in 1991.

In May 1993, Mike Goodkind, a spokesperson for Stanford Medical Center, said that the patient 'was very happy' and has a 'skinny abdomen now'.

The biggest foreign creature that can live inside of the human body is any number of parasitic worms. For example, there are some 800,000 species of nematodes. The largest can grow to several metres in length, but usually inhabits only the placenta of the sperm whale. Theoretically, if a human ate the whale, the worm could infect the human – at least for a while.

Of course, bacteria and viruses live in us too. The largest bacterium known to exist is a recently-discovered, single-celled organism which normally lives in the bowels of a fish inhabiting Lizard Island, Queensland.[1] Until this year, bacteria were thought to be so small that none could be seen with the naked eye. However, this new mammoth bacterium is about the size of a hyphen in a newspaper.

Dr Ester Angert, researcher at the University of Indiana in Bloomington and co-discoverer of the giant bacterium, said that 'it's so huge that we could stick electrodes in it'.

The bacterium is named Epulopiscium fishelsoni and the fish is named Acanthurus nigrofuscus. Again, theoretically, if we ate this fish, a human could host this bacterium. But wanting to stay in us permanently is another matter. Perhaps it would not care to live in something so big.[2]

¶ Why are we getting taller?

It is true that humans have been growing taller throughout history. The increase in the average height of males has been about 10 cm over the past 200 years. Female height is between 6 and 9 per cent below that of males.

This phenomenon appears to have occurred in virtually every country in Europe and in other developed nations. Increasing size has been associated with industrialisation, improved living conditions, increased nourishment, better medical care and improved hygiene and sanitation causing fewer diseases. At least four theories have been put forward to explain our growth:

- || abundant food supplies, especially animal protein, which prevents growth stunting;
- || large increases in the consumption of sugar which stimulates growth;
- || immunisation during childhood which results in fewer growth-inhibiting diseases;
- || less physical labour during the childhood years which allows food energy to be used to build bigger bodies.[3]

¶ How little can a person be?

We are capable of doing things in small ways too. The world's smallest human ever to have lived was Lucia Zarate (1863–1889) of San Carlos, Mexico. She weighed 5.9 kg at age 20. This was the heaviest weight she ever achieved. At birth she weighed only 1.1 kg. At age 17 she weighed only 2.1 kg. Her adult height was only 67.3 cm. She suffered from a particularly devastating form of dwarfism (nanosomia) which is now more commonly called dysplasia. She was unable to grow larger and put on weight.

In fact, there are at least 29 separate forms of dysplasias. One of the more common is achondroplasia, which occurs in about one in every 25,000 births. The basis of the condition is genetic, but its cause is unknown.

Hypophysial infantilism (formerly called ateleiosis) is perhaps the most growth restricting. Individuals with this are usually the shortest people of all. They have essentially normal proportions but suffer from growth hormone deficiency.

It is known that Ms Zarate lived in poverty. She probably was unable to gain adequate nutrition. Although genes and hormones are chiefly involved, poor nutrition contributes to such conditions. Such individuals tended to be shorter in the past due to poorer nutritional standards in historical times.

Although Ms Zarate's body was tiny, evidently her intellectual capacity was normal. This highlights a major myth held about people with dysplasias: a small body means a small brain. In fact, although mental retardation does occur in some dysplasias, in most cases the brain and mind may be fully capable of whatever normal humans can do.

The late actor Michael Dunn is a case in point. Dunn, 'proud to be a dwarf' he used to say, was known as possessing the finest memory for movie lines in Hollywood. He supposedly could memorise an entire Shakespeare play in one reading. A member of Mensa, the intellectual organisation of people with high IQs, Dunn at one stage held the Mensa performance record. Dunn received an Academy Award nomination for his role in *Ship of Fools* (1965) and starred on Broadway despite the few roles for him.

The world's smallest surviving human baby weighed 283.5 gms at birth. A normal full-term newborn weighs about 3,200 gms. About 5 per cent of human newborns weigh more than 4,000 gms, while about 5 per cent weigh under 2,500. Infants weighing less than 2,500 gms are classified as 'low birthweight' and less than 1500 gms as 'very low birthweight'. Even a twenty-week-old foetus weighs an average of 300 gms. At birth, this minutest of humans was only 30.4 cm long. A normal full-term baby is about 48.2 cm long.

The tiny infant was born to Mrs Marian Taggart (1938–83) in South Shields on 6 December 1961. The baby girl was born six weeks premature. Astonishingly, the birth was unattended.

When Mrs Taggart finally received medical treatment for her baby, that attention was nothing short of heroic. Dr D.A. Shearer fed the tiny infant hourly for the first 30 hours with a solution of glucose, brandy, and water – fed through a fountain-pen filler.

The careful, round-the-clock attention paid off. At three weeks, the tiny girl weighed 822 gms. At one year, she was 6.29 kg. By her twenty-first birthday, she weighed a normal 48 kg.

The smallest part of the human body is the human cell. Human cells come in various sizes. The male sex cell, the sperm, is the smallest. This is in contrast to the female sex cell, the ovum, which is the largest – the only one capable of being seen with the naked eye.

The stapes, or stirrup bone, is the smallest human bone. It is one of three auditory ossicle bones in the middle ear, weighs 3.23 milligrams, and measures 0.38 cm. This is in contrast to the largest bone, the femur (thigh bone). The femur comprises about 27.5 per cent of an adult's height.

As for muscles, the smallest is the stapedius which controls the stapes. It is only 0.127 cm long. The stapedius could not be more different in form and function from the body's largest muscle – the gluteus maximus (the muscle of the buttock). In this case, the biggest is always behind.[4]

¶ Can some people predict the weather from the pain in their joints?

Surprisingly, it is possible for people to predict the weather by 'tuning into' their pain. People who have had joint injuries may indeed suffer extra aches and pains when the barometric pressure falls, according to Dr Paul Ort, an orthopaedic surgeon at the New York University Medical Center. However, medical or behavioural science does not presently know why this occurs.[5]

Dr Joseph Hollander of the University of Pennsylvania Medical School argues that a rise in humidity and a decrease in barometric pressure often leads to increases in pain for those suffering from arthritis.

It has been asserted that cell permeability may help to explain the pain. More blood fluid may be forced into the tissues of arthritis sufferers. Blood vessel walls are often more permeable in those with arthritis and blood is always under higher pressure than the surrounding body tissues. This movement would be greatest when the pressure of the surrounding environment outside the body is lowest – as it is just before a storm. If joints are already sore, stiff, swollen, and inflamed, the added fluid could trigger the extra pain. This still unproven theory is perhaps the best science has been able to come up with so far.

Dr Melvin Rosenwater, an orthopaedic surgeon at the

Columbia College of Physicians and Surgeons in New York, is convinced that nerve cells in joints may be sensitive to changes in barometric pressure. He calls this 'the barometric ache'. But he hastens to add that 'it is not a real clinical problem'.[6]

He stresses that barometric ache may be treated with anti-inflammatory medications such as aspirin – or a trip to a different climate.[7]

¶ Can our bodies warn us of earthquakes?

Throughout history there has been the occasional individual who has claimed to have this ability. Although it has never been historically documented, an American Indian by the name of Shotola supposedly warned Enrico Caruso, the famous Italian opera singer, not to remain in San Francisco just before the 1906 quake and fire devastated the city. Shotola claimed that he could 'hear earthquakes' before they happened.

Today, after a major earthquake strikes, there are sometimes those who boast that they knew it was coming. However, such people are rarely believed since they invariably come forward only after the event. The public generally, and scientists in particular, are rightfully dismissive of such claims. In fact, history fails to record a single documented case where an individual accurately predicted an earthquake using only their senses. Thus, it would seem that the human body cannot warn us of earthquakes.

Nevertheless, there is striking evidence that animals may possess some form of earthquake-predicting ability.

Dr Helmut Tributsch argues that animals unquestionably have the ability to predict earthquakes. Dr Tributsch, a German physicist and chemist, contends that the ancient Greeks were the first to recognise this in animals.[8]

He maintains that hours, even days before earth-quakes, distinct changes in the behaviours of animals serve as the tip-off that an earthquake is imminent. Furthermore, he believes that animal behavioural changes can forecast an impending earthquake long before one is predicted using the most sophisticated seismographic equipment available.

Dr Tributsch cites numerous examples of various changes in animal behaviour immediately before an earthquake.

|| Animal unrest

'Animal unrest' occurred in a Peruvian town just prior to a major earthquake. Hens would not return to their coops, mother cats carried their litters out of homes and into open spaces, cows mooed eerily and incessantly, and so on.

|| Fish migrate

Fishermen in Japan have long recognised changes in the migratory patterns of fish right before an earthquake. A 1932 statistical report on mackerel catches near Japan's Izu Peninsula seems to confirm this. It was found that during the 1920s, the annual size of the fish haul varied directly in relationship to earthquake activity.

Furthermore, although sardines are rare in the Pacific, the shallow coastal waters near Miyagi, Japan, were teeming with schools of sardines prior to a major earthquake striking near there in 1933.

Japanese folklore claims that fish foretell earthquakes. Before the Tokyo earthquake of 1855, a fisherman noted a large number of catfish frantically splashing on the surface of the water. Recalling this point of folklore and realising that catfish are normally sluggish, bottom-dwelling fish, the quick-thinking fisherman was able to return home and drag the furniture out of his house just before the earthquake struck.

|| Rats migrate

A restaurant called the House of Rats in Nagoya, Japan, lost its claim to fame when its rats suddenly disappeared. The rats had been allowed to roam free in the restaurant. On 27 October 1891, one day after the rats left, an earthquake struck with an estimated force of 7.9 on the Richter scale. Police in California reported large numbers of rats scurrying through the streets the day before a major earthquake struck the San Fernando Valley on 9 February 1971.

|| Snakes migrate

The 7.3 earthquake which struck the Liaoning Province of China on 4 February 1975 and caused massive destruction ought to have caused more loss of life. However, few people were killed since most had already evacuated the area. They did this because strange animal behaviours began occurring two months prior to the earthquake. Most notably among these behaviours, snakes had come out of winter hibernation to migrate but instead were freezing to death in the snow.

|| Bats fly during the day
Just prior to the 7.0 earthquake which struck Uzbekistan in the USSR in 1976, people noticed masses of bats flying during the daytime. The local government authorities of nearby Turkey now take the appearance of bats in daylight hours as a warning of an impending earthquake.

|| Birds will not roost
Before the major earthquake hit Friuli, Italy, in May 1976, local residents reported seeing many odd animal behaviours. Among these, birds emitted strange calls and nesting birds refused to roost.

|| Cattle head for high ground
In March, 1964, a grazier on Kodiak Island, Alaska, decided to head for higher ground. He did so following a hunch he had based on the behaviour of his cattle. After herding his cattle in the same way for years, one day the cattle suddenly stopped grazing and just as suddenly started heading for higher ground. The grazier was puzzled and thought he might just as well see why they

were behaving this way. Two hours later, a major earthquake struck which devastated the city of Anchorage. This was followed by a huge tidal wave which flooded the lowlands which the cattle and the grazier had just occupied.

|| Chimpanzees get hyperactive

At the Primate Study Center of Stanford University in California, researchers routinely track the behaviour of chimpanzees. In 1975, it was noted that there was a far greater likelihood that an earthquake occurred on the day after a day in which the chimps were the most active. In addition, after the Loma Prieta earthquake near San Francisco, in October 1989, 'more anecdotes made the rounds' of strange animal behaviour prior to the quake. For example, 'at the aquarium in Monterey (south of the epicentre), two divers in a tank noticed that fish stopped swimming in formation just before the quake – and returned to formation afterward'.[9]

Dr William Bakun, chief seismologist for the US Geological Survey in Menlo Park, California, admits that 'there are ample reports worldwide of animals going haywire prior to earthquakes'.[10]

¶ Why does animal behaviour change?

As to what it is about earthquakes that alters the behaviour of animals, Dr Tributsch speculates that it is probably a combination of factors. He suggests several possibilities: 'Venting ground gasses could excite animals ... perhaps, changes in the groundwater table.' Moreover, he adds that animals may sense 'changes in atmospheric pressure, slight swaying of the ground – even ultrasound emitted from rock that is nearly bursting'.

Dr James Berkland, a geologist in Santa Clara, California, contends that changes in the earth's magnetic field 'disorient' animals and change their behaviour. He believes that such magnetic field changes occur prior to earthquakes.

This controversial view is frowned upon by many scientists. However, it has recently been bolstered by perhaps a coincidental event. According to Dr Anthony Fraser-Smith, an atmospheric researcher at Stanford University, the Stanford magetometer recorded 'highly unusual changes' just prior to the Loma Prieta earthquake. The magnetometer usually only detects magnetic field variations due to solar magnetic storms and other effects high in the earth's atmosphere.

Could we ever rely on animals for earthquake prediction? Dr Bakun does not dismiss this possibility. He claims that perhaps 'if we understood what these creatures were responding to, we could build an instrument to measure that phenomenon'.

Dr Berkland has a simpler strategy. Since one of the pre-earthquake behaviours is that more pets run away from home, he thinks that scientists should monitor carefully the lost-and-found ads in newspapers.[11]

If animals can predict earthquakes, could humans ever learn such a skill from them? Theoretically, such a possibility exists, but it seems highly improbable.

Still, if someone comes up to you and says that he hears earthquakes – it doesn't hurt to listen.[12]

¶ What crippled Tiny Tim?

In 1972, research seemed to have solved this recurring Christmas medical mystery: what was wrong with Tiny Tim? Now, we're not so sure.

Tiny Tim, the Cratchit family's crippled little boy in *A Christmas Carol* (1843), is one of Charles Dickens' (1812–1870) most beloved characters. It is often assumed that Tiny Tim suffered from polio. But this was probably not the case.

Tiny Tim's problem was the subject of an insightful and imaginative article appearing in the *Australian Paediatric Journal*.[13] Its author, Dr Peter Jones, is now retired from his post at the Royal Children's Hospital in Melbourne. According to Dr Jones, only 'sparse information' exists upon which one can make a reasonable medical diagnosis. We only know from Dickens' description that Tiny Tim 'bore a little crutch and had his limbs supported by an iron frame'. Furthermore, he was carried around on the back of his father, Bob Cratchit.

Although this sounds very much like a child with polio, Dr Jones points out that 'tuberculosis of the hip is a far more likely diagnosis; "coxalgia", as it was known from its presenting symptom, carried with it a sentence of gross deformity and frequently death...'.

He notes that polio is an unlikely diagnosis since 'poliomyelitis is paradoxically more prevalent in communities with high standards of municipal and domestic hygiene – and no-one would suggest that this could be said of London in 1843'.

As for Tiny Tim's chances of recovery, Dr Jones prefers to 'carry optimism to the extreme as Dickens did' and suggest that Tiny Tim would survive. He adds that perhaps Tiny Tim might have had 'pseudocoxalgia' or Perthes' disease, a form of osteochondrosis (although this disease was not defined for another 77 years).

However, a US doctor claims that Tiny Tim did not suffer from a bone disease at all. Instead, he had a kidney disease that made his blood extremely acidic and eventually resulted in his being crippled.[14]

Dr Donald Lewis, a professor of paediatrics and neurology at the Medical College of Hampton Roads in Norfolk, Virginia, studies Tiny Tim to show medical students how to diagnose children. Dr Lewis has his students read Dickens and search for symptoms.

According to Dr Lewis, Dickens could not have known about the kidney condition called DRTA-1 (Distal Renal Tubular Acidosis Type 1) because the disease was not recognised until the early twentieth century. Nevertheless, therapies for symptoms of this disease were used in Dickens' day and might have successfully treated it.

Dr Lewis argues that Tiny Tim had classic symptoms of DRTA-1. In addition, Dr Lewis believes that Tiny Tim may have also suffered from neurological problems due to Tiny Tim's spells of weakness, withered hand and limpness.

In any case, so great was the impact of *A Christmas Carol* on nineteenth-century England that a charitable trust for crippled children was founded – the Tiny Tim Guild.

God bless us every one![15]

¶ Why do feet swell up in aeroplanes?

It is a myth that feet swell up when you ride in an aeroplane because of changes in atmospheric pressure due to high elevation. Feet swell up on planes, especially during long flights, for the same reason they swell up on the ground – inactivity.

The heart is not the only body organ that serves as a pump – leg muscles do too. Walking or flexing a leg assists in the pumping effect of the heart. If, while on a plane, you are not only confined but also sitting, gravity forces blood and other fluids to the lowest body point – the feet.

The 'pooling' of body fluids in the feet can happen just as easily in a bus, train, car or office. In fact, most people's feet normally swell somewhat during the day in any case – some up to a full size larger.

If you remain inactive while on a long flight, it does not matter if you leave your shoes on or off. They will swell either way. If left on, they will provide external support, but will inhibit circulation a bit more and probably feel tighter during the latter part of the flight. If taken off, comfort is probably increased, but shoes are likely to be more difficult to put on once the flight is over.

Podiatrists normally recommend aeroplane aerobics to help circulation – including help for swelling feet.

¶ Can a person be identified by DNA from a bit of bone?

It is now well known that scientists can identify a missing person from just a tiny fragment of bone, a single tooth or a bit of DNA. Skeletal remains, bones and teeth can reveal the age and height at death, as well as sex and even ethnicity.

According to Drs Christopher Joyce and Eric Stover there are some useful keys that forensic anthropologists use to unlock the identification door.[16]

‖ Age
When scientists examine skeletal remains and want to know the age of the person at the time of death, they first look to the back of the skull. The occipital bone at the base of the skull fuses and hardens at age four or five. However, eight other bones continue to grow as a person gets older. Thus, considered together, the condition of these bones can establish an age range of up to 40 years. Next, scientists look to the teeth. The pattern of teeth appearance (eruption) is also useful in determining age. For example, the dental remains may consist of only deciduous (baby) teeth, only permanent teeth, or a combination of both.

|| Sex

When the gender of a person is sought, scientists begin by examining the pelvic region. A woman's pelvic bone opening is wider than a man's. This better accommodates childbirth. Next, the front of the skull is examined. The ridge of the brow is almost always more prominent in men than in women.

|| Height

When the height is sought, scientists first look to the thigh. The length of the femur (the long bone of the thigh) is directly related to a person's overall height. Thus, if the femur is available, or even just enough of it so that the femur's length can be calculated, then the entire height of the person can also be calculated. For this last step, forensic anthropologists use special, anthropometric tables.

|| Ethnicity

When scientists want to determine ethnicity, they again look to the skull. Forensic anthropologists claim to be able to successfully identify ethnicity in about 90 per cent of cases by measuring a number of subtle skull differences.

|| Medical and dental records

Although by themselves skeletal remains give forensic

anthropologists valuable evidence, skeletal data alongside medical or dental records yield even greater information. When an unidentified body is found, scientists make observations and conduct tests of the skeletal remains. They then compare observations and tests results with the medical or dental records of a 'suspect'. In addition to the examinations for age, sex, height and ethnicity, there are three other means of establishing identity when both skeletal evidence and medical records are available.

First, the history of one's dental work is documented in dental records. The number of teeth, bridge work, and other aspects of dental composition may in itself lead to a positive identification. Teeth are coated with tough enamel. As such, they are more resistant to deterioration compared to other parts of the skeleton. Each tooth is also unique in the shape of its crown, roots, and pulp cavity.

Second, injuries to bones or teeth often leave permanent marks. These too can be documented in medical or dental records.

Third, as with teeth, bones are unique in their size, shape and the details of their internal structure. These elements are readily revealed by x-ray. The x-rays are then compared to those in the medical records. According to Dr Clyde Snow, a University of Oklahoma forensic anthropologist, 'bones are just as unique as fingerprints'.

Thus, due to this uniqueness, it is possible to identify a person from a fragment of bone or just one tooth.[17]

But what if there are no available medical or dental records or if only a few fragments of bone exist? Advances in research now make it possible to extract genetic material – DNA – from the bone and teeth fragments of people even if long dead. Identity can be established by comparing these samples with samples from living relatives. In this process, a particular DNA, called mitochondrial DNA, is especially useful.

In every human cell, there are coffee-bean-shaped structures called mitochondria located outside of the cell nucleus, but well within the cell wall. The mitochondria generate energy for the cell by breaking down sugars and fats. Within the mitochondria, DNA is found in great abundance – often in amounts hundreds of times greater than in other parts of the cell. Experience has shown that if any DNA survives it is most likely to be this DNA of the mitochondria. Since mitochondrial DNA is passed only from a mother to all of her offspring, a person's mitochondrial DNA is identical to that of their mother, sisters or brothers. Barring a rare genetic mutation, identification of a person is nearly always possible.

Thus, in order to establish a missing person's identity, scientists move up and down a family's tree – using DNA as the ladder.[18]

¶ Why are Pygmies little?

African Pygmies have fascinated Westerners since the first photographs of these miniature people reached Europe more than a century ago. But Pygmies have also baffled scientists. The normal explanation for short stature simply does not apply to Pygmies.

Short stature is usually caused by low levels of human growth hormone (HGH). This hormone is secreted during the night by the pituitary gland located in the brain. Shortness (or tallness) runs in families because the amount of HGH an individual possesses is genetically determined. The gene for HGH is one of a cluster of five closely related genes located on chromosome 17. The hormone itself is actually a complicated protein made up of 191 amino acids.

Growth will be reduced if the pituitary fails to secrete enough HGH, if a person is afflicted with a genetic abnormality involving chromosome 17 or if HGH is adequate, but for some reason cannot be put to use by the body.

Growth can also be stunted by nutritional deficiency, illness and injury – especially as seen throughout the starving Third World.

Any abnormally undersized person is medically termed a dwarf. They exhibit a condition known as nanosomia, which is sometimes called nanism. An artificial, genetically engineered, biosynthetic HGH has been available in some countries for years to treat nanosomia. According to Dr G.L. Warne of the Royal Children's Hospital in Melbourne, biosynthetic 'growth hormone sounds like a miracle drug, and in many ways it is. It has so far not been associated with any serious side effects.' However, Dr Warne points out, 'biosynthetic HGH is extremely expensive ... The course of treatment, involving daily subcutaneous injections, can last for many years.' [19]

For nearly 40 years scientists have known that Pygmies have normal amounts of HGH. And compared to their taller African neighbours, Pygmies partake of an adequate diet, enjoy reasonable health and life expectancy, and do not experience abnormal rates of injuries. So why is it that Pygmies rarely grow taller than 140 cm?

The most recent scientific thinking suggests that Pygmies are short because of a 'cell receptor' failure in their bodies. Thus, although they have sufficient amounts of HGH, they fail to process it correctly.

A cell receptor normally provides an all-important 'docking site' for the HGH circulating in our bloodstream. The HGH must 'dock' with a cell in order to be utilised by that cell. But the cell receptor is somehow deficient in Pygmies and does not allow enough docking to go on. Hence, this cell receptor failure causes much of the HGH to be wasted.

Drs Gerhard Baumann and Melissa Shaw of Northwestern University Medical School in Chicago, along with Dr Thomas Merimee of the University of Florida in Gainesville, analysed the blood of 20 Pygmies living in the Ituri Forest of central Zaire and came to some interesting conclusions.[20]

The Baumann team found 'debris resulting from cell receptor failure' within the Pygmy blood. This debris consisted of molecules of HGH attached to a binding protein broken off during an unsuccessful docking attempt.

The Baumann team discovered that Pygmies have half the amount of the necessary hormone-protein complex compared with a group of individuals serving as control subjects. Thus, this low level of the hormone-protein complex 'strongly indicates' that Pygmies do in fact have a shortage of HGH cell receptors. With so few receptors making docking difficult, Pygmies almost certainly would have problems processing HGH – even if they received it artificially.

The theory put forward by the Baumann team is consistent with the often-observed characteristics of Pygmy growth. Pygmies show seriously stunted growth during adolescence. This is precisely when a full comple-ment of cell receptors is most essential for normal growth. In fact, normal Western teenagers usually produce extra HGH cell receptors during this period – thus allowing the typical adolescent growth spurt.

Intriguingly, the Ituri Forest Pygmies demonstrate that, although they are short, they are certainly extremely able-bodied. Both men and women are exceedingly agile, nimble and strong.

But it is their children that pose perhaps the greatest fascination. According to Jean Pierre Hallett in both the book *Pygmy Kitabu* (1973) and the film *Pygmies* (1974), Pygmy children from a very young age have extraordinary physical abilities.[21] For example, three-year-old Pygmies can accurately shoot an arrow – hitting a 6 cm wide target nine out of 10 times at a distance of 10 metres. They can use a sling to down small prey, swim confidently, and run faster than the average Western child at age eight. At three, the typical Western child is scarcely able to balance on one foot, while a three-year-old Pygmy child can effortlessly scale a 20-metre-high coconut tree!

Actions speak louder than size.[22]

¶ Do we really have a 'funny bone'?

We do not really have a funny bone – it is really a funny nerve. It is a nerve called the ulnar nerve and it supplies sensation to the arm, hand and fingers. For most of its length, the ulnar nerve is located deep under the skin where it is well protected.[23] However, at the elbow it comes very close to the surface and is covered only by skin and a thin layer of connective tissue. This is why it hurts so strangely when you bang your elbow in a certain way. You have actually momentarily traumatised your ulnar nerve. The pain sensation is over in a few seconds. It is ironic that someone named it 'funny'.[24]

¶ What is 'writer's cramp'?

Writer's cramp is actually a localised muscle spasm called focal dystonia. It is caused by holding a pen or pencil too long, especially if held too tightly. Relaxing the hand periodically, exercising the hand, holding the pen more loosely and taking frequent breaks from writing usually solve the problem. A way to test if you are holding the pen loosely enough is if your non-writing hand can easily pull the pen from your writing hand. A physician or other health professional would recommend exercises and relaxation techniques that help overcome writer's cramp. However, if a serious injury results, surgery is done – but only in the most severe cases.[25]

Drs Lee Tempel and Joel Perlmutter from Washington University in St Louis studied blood flow in patients with writer's cramp and compared them to people without such symptoms. They found that, on average, the patients with writer's cramp showed only two-thirds as much blood flow in the brain's sensorimotor cortex. This is the brain region responsible for hand sensation and movement.[26]

¶ What causes a 'stitch' in my ribs?

The pain is really not in your ribs. A 'stitch', 'runner's cramp' or 'sideache' as it is variously called, is indeed a cramp. A cramp is a painful spasmodic muscular contraction, especially a tonic (tension) spasm. But a 'stitch' actually occurs in the intestines.

As Dr Averil Ma, a gastroenterologist at the Columbia Presbyterian Medical Center in New York, explains, 'Several things can cause such a cramp. For example, if someone ate a heavy meal and then started doing something else that demands blood flow, there would be relative ischemia (a reduction in the blood supply to the intestines). This would produce a cramp. Muscular problems can also cause such pain.'[27]

¶ Do people who have lost a limb still sometimes feel sensation in it?

This refers to the fascinating phenomenon of Phantom Limb Syndrome.

Sensation in lost arms and legs reported by amputees has intrigued physicians throughout the centuries. Phantom Limb Syndrome has been scientifically studied with reports in the medical and behavioural science literature since at least 1940.

According to Dr Ronald Melzack of the Department of Psychology at McGill University in Montreal, writing in *Scientific American*, Phantom Limb Syndrome (PLS) occurs in up to 70 per cent of amputees. The sensation is often painful and has been described as burning, cramping or shooting. It can vary from occasional and mild to continuous and severe. It usually starts soon after the amputation, but sometimes appears week, months or years later.

Dr Melzack adds that 'the oldest explanation for phantom limbs and their associated pain is that the remaining nerves in the stump, which grow at the cut end into nodules called neuromas, continue to generate impulses'.[28]

¶ Why do women have one more rib than men?

Sometimes men have something extra. In this case, women often do. Even a simple observation of skeletons leads to the conclusion that women have an extra rib more often than men. It probably makes no difference in human health.

Some of us have fewer bones than others within and between genders. A baby is born with about 350 soft structures that eventually 'fuse' in to the 208 solid bones of adulthood. The fusing process is slightly different in all of us.

Compared with men, women have a greater likelihood of having an extra rib or two probably due to variation in the fusing process.

The Bible story is much more romantic. According to Genesis, God took Adam's rib while he was asleep, gathered some mud and fashioned Eve. (And you thought little girls were made from sugar and spice and everything nice!). In taking the rib from Adam, God performed the first surgery on a human.

'Adam's Rib' Syndrome or, more technically, Thoracic Outlet Syndrome (TOS) is caused by a problem rib. Often it is an extra rib that the body does not need or one that has moved into the wrong place through injury or other causes. It occurs in both women and men. If you suffer from TOS, as you move your head or arm this problem rib pushes against the arteries, veins and nerves that travel to your arm. This causes a numbing sensation. An almost certain sign of this disorder is when the pulse of your arm stops whenever you move your arm or neck.

If intensive physical therapy and exercise do not help, then surgery may be needed. It is a rather delicate operation.[29]

Of course, some people have an extra rib removed for cosmetic reasons. For example, the actress Raquel Welch allegedly had ribs removed to help her keep her hourglass figure.

¶ Why do people 'shrink' when they get older?

Seniors do not 'shrink' as such. However, elderly people do experience a loss in their body height. What happens is due less to genes and more to the effects of gravity and time. By age 70, the average man of average weight and height loses about 3 cm from his maximum height achieved during early adulthood. According to Dr Lawrence Riggs of the Mayo Clinic in Minneapolis, three factors are involved:

|| loss of space between the spinal discs due to the pressure of gravity;
|| a general weakening of the back muscles;
|| poor posture.

Women suffer a slightly greater loss in proportion to their earlier height. This is believed to be due to their higher rates of osteoporosis.[30]

¶ Why am I taller in the morning?

All of us are taller in the morning, shorter in the afternoon and shortest of all at night.

Dr Jerry Wales of the Department of Paediatrics at the University of Sheffield explains that there are two components in this. First, in the growing child, 'growth hormone is secreted in pulses overnight. This acts through several intermediary steps to cause lengthening of the bones at the end-plates (epiphyses).' Second, after growth ceases, there is a daily 'postural compression of the spine under the effect of gravity'. The difference in adults is an average of roughly 15 mm from morning to night.

Dr Peter Dangerfield of the Department of Anatomy at the University of Liverpool adds that another component must be considered – the existing curves of the spinal column itself. 'These curves vary with body weight and position. As a result, the spinal column tends to press downwards when in an upright position, altering these curvatures, and hence shortening the spinal length. When lying down, the reverse happens and the column lengthens again. It is estimated that 80 per cent of the height change is accounted for by changes in these curvatures.'[31]

9 · **The Inside**

The ancient Greek playwright Sophocles (495–406 BC) once wrote, 'Wonders are many, and none is more wonderful than man.' Yes, humans are wonderful. The deeper we probe into this wonderful organism, the more we reveal how wonderful we are.

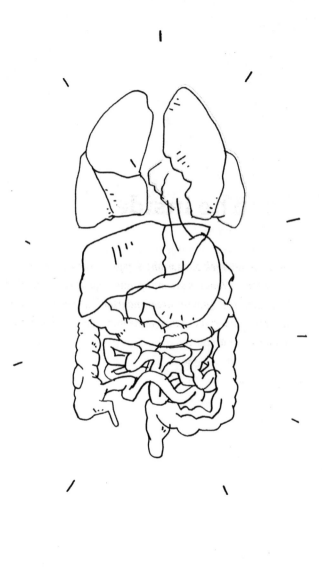

¶ What is pain?

We all feel it, but what is it really?

Pain is defined medically as a more or less localised sensation of discomfort, distress or agony resulting from the stimulation of specialised nerve endings. However, research shows that pain is far more complicated.

Pain is necessary for our survival. It serves as a protective mechanism – the body's danger alarm – insofar as it warns us to remove ourselves from the possibly hazardous or even fatal pain source. For example, when we burn our fingers with a match, the body is telling us, 'Do not expose me to this high temperature any longer or else permanent cell damage will result.'

Medical science distinguishes at least 36 separate kinds of pain. 'Bearing-down pain' accompanies uterine contractions during the second stage of childbirth labour. 'Growing pains' are recurrent quasi-rheumatic limb pains peculiar to adolescence. Sharp, darting pains are called 'lancinating pains'.

Some pains are bizarre and difficult to explain. For example, 'phantom limb pain' is felt as arising from a limb, although that limb has been amputated.

Some pains are named after people. 'Brodie's pain' is pain induced by folding the skin near a joint affected with neuralgia. 'Charcot's pain' results from rheumatism of a testicle.

Research shows that pain is not merely a simple electrical nerve response. Perhaps surprisingly, there is a relative independence of pain from the body's own pathological processes. Many studies show that pain may be reduced with a placebo (a dummy medical treatment). It also can be psychologically induced through the use of a nocebo (a dangerous medical treatment). This has been demonstrated many times.[1]

Pain stimuli may also evoke differing behavioural responses according to age, prior psychological status, type of group the person belongs to (for example, religious group), social context and culture.

The amount of pain inferred or attributed to nine different conditions is 'culturally learned'. This is according to two University of Alberta psychologists, Drs Janice and Robert Morse, who studied subjects from four different cultural groups residing in western Canada: Hutterites, East Indians, Ukrainians and Canadian Anglophones. The study found that each of the four groups rated and reacted to each of the nine different pain stimuli differently.[2]

Thus, pain is not just your nerves speaking, but your experience too.[3]

¶ What is a laugh?

Physiologically, laughing is a series of spasmodic and
partly involuntary expirations with odd vocalisations,
normally indicative of merriment. Often, laughing is
an hysterical manifestation or a reflex result of tickling.
Normal laughing is of two types: mild and hearty, for
which there is an occasion. But abnormal laughing is
of three types: compulsive, forced and obsessive, for
which there is no occasion.

¶ What makes me laugh?

Other than simple tickling, laughing is based on fear. Fear of social embarrassment, of loss of dignity, of exclusion from a group, of being fooled or exploited, of death, of injury, of sex. There is a fine line between comedy and tragedy, between what makes us laugh and what makes us cry, between pleasure and pain.

¶ What happens to my body when I laugh?

When you give way to laughter, electrical impulses are triggered by nerves in your brain. These set off chemical reactions in the brain and elsewhere in the body. For example, your endocrine system orders your brain to secrete natural tranquillisers and painkillers. Other released chemicals aid digestion. Still others make arteries contract and relax and improve blood flow. Laughing may not be the best medicine, but it's certainly a good one.

¶ Why is laughing important?

Among other things, laughing restores balance and equilibrium. Charles Darwin in *The Expression of the Emotions in Man and Animals* (1872) argued that laughing helps us discharge surplus tension and mental excitation. Sigmund Freud argued that laughter helps us deal with lustful thoughts. Laughing is important to our very survival. Laughing starts when we are about 12 weeks old. Darwin argued that a baby laughing gives pleasure to the caretaker and thus helps lessen the likelihood of parental rejection – both aiding personal and species survival.

¶ Does laughing keep us healthy or even make us well?

Dr Norman Cousins, in his book, *Anatomy of an Illness as Perceived by the Patient* (1979), claims that laughter helped him regain his health.[4] His evidence was so overwhelming that some hospitals now have 'laughing libraries' of videos to help patients laugh their way to recovery, and there are also 'clown wards' in hospitals – especially children's wards.

Biochemically, laughter reduces the body's production of cortisol. It is known that cortisol suppresses the body's immune system. Thus, by laughter, the body's immune system is left unimpeded by cortisol.

It is said that 'a laugh a day keeps the doctor away'. Is it true that laughter has curative powers?

Research shows that when we laugh, our metabolism rate picks up, muscles are massaged and stimulated and a variety of biochemical substances rush into the bloodstream. Studies demonstrate that after a period of laughing, subjects not only feel momentarily relaxed, but they also have fortified themselves against depression and heart disease, and heightened their resistance to pain. Now researchers think that laughter may boost the immune system as well.

This has been shown in several experiments. For example, in one experiment conducted by Dr Kathleen

Dillon at the Western New England College in Springfield, Massachusetts, university student volunteers were divided into two groups. One group was instructed to watch a non-humorous educational video. The other group watched a humorous video of Richard Pryor comedy routines. Dr Dillon found that concentrations of salivary immunoglobulin A (IgA) – an antibody linked to lower rates of upper respiratory illness – measurably jumped in the 'humour video' group.[5]

According to Dr Lee S. Berk, an immunologist at the Loma Linda University School of Medicine in California, 'negative emotions can manipulate the immune system, and it now seems positive ones can do something similar'. Although the subject of much speculation, Dr Berk believes that laughter starts a simple biochemical process in motion involving the body-produced chemical, cortisol. He says, 'cortisol, which is an immune suppressor, has a tremendous influence on the immune system. Laughter decreases cortisol, which allows interleukin–2 and other immune boosters to express themselves.'

The possibility of the curative powers of laughter has already begun to change the practice of one major hospital – at least experimentally. Columbia Presbyterian Medical Center in New York City has established 'The Big Apple Circus/Clown Care Unit'. In this facility, laughter is continually emphasised. For example, professional clowns are employed to dress in white hospital coats and race around the ward on roller skates. Practical jokes and pranks are constant. 'Let's draw your blood', says a 'staff' member, and a crayon and sketch pad are produced. So far, jokes that bomb have not resulted in malpractice suits.

No-one knows for sure if 'a laugh a day keeps the doctor away'. But Stephen Sondheim's classic song takes on a new meaning: illness may be the time to 'send in the clowns'.[6]

¶ Does laughing make us more productive?

Research shows that industrial productivity is boosted by humour.

Evidence suggests that humour in the workplace is a powerful profit-making tool, and some of the world's largest corporations such as IBM, Monsanto and General Foods are cashing in.

At the University of Maryland, Dr Alice Isen's studies show that 'people put in a good mood organise information better and are more creative'. Furthermore, 'people in good spirits proved more creative in word association, categorisation and tasks involving memory'. Moreover, 'humour also improves decision making and negotiating abilities'.[7]

Dr Isen says that 'mild elation seems to lead to the kind of thinking that enables people to solve problems requiring ingenuity or innovation'. She adds, 'someone who is happy can perceive subtle relationships between things because positive material is stimulated in his memory. He has more ideas.'

One of Dr Isen's experiments produced some intriguing results. Adult volunteers were divided into pairs playing fictitious roles of buyers and sellers of goods. They had to successfully perceive a range of alternatives and combine them cleverly in order to achieve the highest profits or best deal. Interestingly, the pairs who viewed humorous cartoons beforehand were less contentious and 'likelier to reach solutions that were mutually beneficial'.

Dr Isen says that 'combining issues and developing novel solutions may be necessary for anything that goes beyond obvious compromises. These are the very same capabilities that are enhanced by positive feelings.'

At the University of Tennessee, Dr Howard Pollio's work shows that 'humour improves group as well as individual performance'. In fact, 'when it's related to the task at hand, laughter tends to boost performance'. He asserts that laughter provides a 'brief break without being too diverting', relieves boredom and encourages problem solving.

At the California State University–Long Beach, Dr David Abramis recently surveyed 341 adults and found that they 'take fun at work very seriously' and consider 'a sense of accomplishment to be "fun"'.

Interestingly, Dr Abramis has discovered a 'clear relationship between intending to have fun and actually having some: those who think fun belongs at work enjoy themselves most'. So the humour research message is clear: 'lighten up a little and laugh it up a lot. It can only do you – and your employer – a whirl of good.'

Related research indicates that sadness, lack of humour, and pessimism may detract from one's employment performance and may even negatively affect one's health and life-expectancy.

A monumental study spanning 35 years has followed graduates of Harvard University through their careers. Dr Christopher Peterson of the University of Michigan and two colleagues reported in the *Journal of Personality and Social Psychology* on this longitudinal study of the attitudes and health of the elite.[8]

The Peterson team found that there exists a clear relationship between an 'optimistic outlook' and success. And correspondingly, there is a relationship between a 'pessimistic outlook' and lack of success in career accomplishment. In fact, even more profound is that:

> individuals who explain bad events pessimistically in early adulthood (at the time they were first surveyed at Harvard as recent graduates just after age 25) have substantially more illness at age 45 than those who offer rosier explanations for bad events. The relationship between pessimism and poor health declines somewhat in the following years but remains statistically significant through to age 60.

Evidently, this 'good humour' research is being taken very seriously by corporations. The personnel director of the New York office of one multinational says that 'we now try to hire MBA grads who are 'up' as well as 'bright'. Who needs execs who get pensioned off at 50? I suppose you could say that we want the best, the brightest and the funniest.'

As of January 1995, there were more than eighty articles in the medical literature attesting to the fact that having a good sense of humour is a factor in getting well and staying that way.[9]

¶ Does laughing make me cope better?

Is simply having a good laugh the best way to cope with tough times? Perhaps so for most of us. But how far can you take the notion that humour fights depression? Research has established that humour has important educational and psychological uses.

Humour can be used as an educational tool to increase rates of learning. This was demonstrated in two famous experiments by Dr Avner Ziv of Tel Aviv University published in the *Journal of Experimental Education*.[10] In the first experiment, Dr Ziv divided 161 university students doing a one-semester statistics course into two groups. One group was taught statistics through the use of humorous teaching materials, while the second group was taught the identical subject matter through the use of traditional, non-humorous materials. At the end of the semester, both groups were tested to see which group, if any, learned the subject matter better. Dr Ziv writes, 'the results showed significant differences between the two groups in favour of the group learning with humour'.

Humour can also be used as a therapeutic tool in clinical psychology in several ways. Summarising these uses in an article in *Psychological Reports*[11], Dr Sharon Dimmer from Michigan State University along with two colleagues wrote that 'humour in psychotherapy can be used to alleviate anxiety and tension, encourage insight, increase motivation, create an atmosphere of closeness and equality between therapist and client, expose absurd beliefs, develop a sense of proportion to one's importance in life situations and facilitate emotional catharsis'.

But despite these beneficial uses, is humour an effective weapon against depression? On this, authorities are not so sure.

On the one hand, we all know that laughing makes us feel good – at least for the moment. Furthermore, many psychology textbooks and clinical therapeutic handbooks recommend the use of humour and laughter in both the diagnosis and treatment of psychological disorders, including depression. For example, in the *Handbook of Humor in Psychotherapy* edited by W. Fry and W. Salameh, it is claimed that humour followed by laughter is incompatible with depression and thus comprises an extremely valuable treatment weapon against it.[12]

¶ What comes first in laughing: the smile on the outside or the pleasing feeling on the inside?

Most people, including many behavioural scientists, would confidently say that the feeling produces the smile, and not vice versa. But if the latest research findings are right, then these people are in for a surprise.

Laboratory experiments suggest that the physiology of smiles and other facial expressions may themselves cause emotions in their own right. This does not mean that facial expressions are more important than thoughts or memories in prompting emotions, only that both interact in ways we previously thought improbable.

Surprisingly too, this idea is not really new. More than a century ago, Charles Darwin and William James (1842–1910), a famous psychologist, both put forward the theory that facial expressions play an important role in bringing about the feelings that accompany them. 'Wipe that look off your face,' said Darwin, 'and you mute the inner feeling.' However, over the years, the Darwin–James theory fell out of favour among behavioural scientists. This remained so until very recently.

Modern research into what is now called 'facial feed-back' began in 1984. In that year, psychologists led by Dr Paul Ekman from the University of California in San Francisco found that when people mimic different emotional expressions their bodies produce distinctive physiological patterns. These patterns include changes in heart and respiration rate, changes in skin temperature and conductivity and changes in general muscular tension level which differ from one emotion to the next.[13]

Dr Ekman was the first to show through various laboratory experiments that people actually feel happier when they smile and feel sadder when their faces are arranged in what he calls a 'configuration of sadness'.

In a more recent study, researchers at Clark University, led by psychologist Dr James Laird, confirmed the finding that getting people to place the muscles of their face in the pattern of a given emotional expression actually produced that feeling. In several experiments, subjects' expressions representing happiness, sadness, anger and disgust all elicited the moods they portrayed.

In another study, University of Michigan researchers had subjects repeat vowel sounds over and over. When subjects pronounced either a long 'e', which forces a mild smile, or an 'ah', which imitates an expression of surprise, pleasant feelings were reported. But when subjects pronounced a long 'u' or an umlauted German 'ü', which both force a mild frown, the opposite emotions were reported – unpleasant feelings.[14]

A leading proponent of the rapidly re-emerging Darwin–James theory is University of Michigan psychologist, Dr Robert Zajonc. Dr Zajonc's studies show that when facial muscles relax they raise the temperature of the blood flowing to the brain, but when the muscles tighten they lower the temperature. He theorises that these temperature changes affect the activity of brain centres that regulate emotion. However, the real impact of these very subtle temperature changes remains very controversial.

Said Dr Zajonc: 'I'm not saying that all moods are due to changes in the muscles of the face, only that facial action leads to changes in mood.'

Dr Laird goes even further. He believes that facial expression not only alters mood, but even alters memory. He says: 'if you're in a good mood, you remember positive things better than negative things. The effect is very specific: when you're in an angry mood, you remember angry things but not necessarily sad things.'

He adds that: 'one way to manipulate mood is to manipulate facial expression. I had subjects read a Woody Allen story and an anger-provoking editorial. When I asked them to smile and recall what they'd read, they remembered more about the Woody Allen story. When I asked them to frown and recall what they'd read, they remembered the editorial better. Produce the face that matches the mood of the content and you'll remember more.'

Precisely what physiological mechanisms are involved and how important is this phenomenon to emotional life?

The next research phase will address these questions and hopefully achieve some answers. Until then, the Mona Lisa's secret is still safe.

There are still many mysteries in a smile.[15]

¶ Why do I laugh when I breathe 'laughing gas'?

'Laughing gas' is the common name for nitrous oxide. This gas was discovered in 1772 by the British scientist Joseph Priestley (1733–1804). Priestley was also the co-discoverer of oxygen. Inhaling nitrous oxide was found to produce a moderate euphoria. This light-headed giddiness was a novelty at fashionable European parties of the nineteenth century. Although Humphry Davy (1778–1829) noticed that nitrous oxide blocked pain and thus might be useful in surgery as an anaesthetic, it was only in the 1840s that nitrous oxide was applied in medicine. Prior to then, all sorts of substances were used as painkillers: alcohol, opium and mandrake to name a few. Today, nitrous oxide is still used, particularly in dentistry.

Nitrous oxide is but one of many gases that can function as an anaesthetic. It works by putting the nervous system more or less to sleep, so to speak, and thereby more or less obliterating consciousness and thus 'killing pain'. Nitrous oxide reduces the oxygen carrying capacity of the blood. From this lack of oxygen we feel 'giddy' and 'high' (hypoxia).

¶ Why can't I tickle myself?

Tickling is one of the least understood of all human physiological reactions. The 'tickle response' is involuntary by definition. Although a person can sometimes control the 'tickle response' by concentrating very hard, it cannot be self-initiated.

It has been suggested that tickling with gentle movements of the fingertips excites certain small, fine nerve endings or touch sensors just beneath the surface of the skin. These are located all over the body, but especially on the palms and soles. The most obvious and observable reaction to tickling is laughter. However, the pulse quickens, the blood pressure rises and the body becomes keyed up and alert as well.

According to Dr Roger Grief, professor emeritus of physiology and biophysics at Cornell University Medical College in New York, the fact that we cannot tickle ourselves is only one of a number of odd things about tickling. Another is that the tickle response is 'ambivalent'. Although the first reaction to tickling is usually pleasure, sometimes this pleasure becomes tinged with anxiety. Hence, the familiar expression 'tickled to death' reflects something of the mingled fear and delight that tickling evokes.

According to Dr William Fry, professor of clinical psychiatry at the Stanford University Medical School,

'If there is no anxiety or danger, people do not laugh or giggle when tickled. Nor will they laugh if they are tickled aggressively enough that they sense danger. People will laugh and giggle if they feel some anxiety but no danger.'

If we tried to tickle ourselves, we would know that at any moment we could cease the stimulation, thus eliminating an essential component of tickling – our anxiety.[16]

¶ Why doesn't my belly button heal over?

This is one of the classic OBQs of all time. Isn't it easy to imagine this being asked by Neanderthals 50,000 years ago? The ancient Greeks? Medieval mystics? Himalayan monks? Kindergarten children – and other great philosophers? The answer is probably not worth all the contemplation.

A belly button, or navel, is merely scar tissue of the umbilical cord where the cord has detached following birth. It is of no medical significance, therefore medical and anatomy texts pay it little or no attention.

Nevertheless, any animal that has been nourished in the womb must have a belly button, although it may not always be easily seen. According to Dr Edward Feldman, a professor of animal reproduction in the School of Veterinary Medicine at the University of California–Davis, 'the scarring may be less obvious in some animals than in others, especially if it is covered by fur'. The belly button does not heal over because there is nothing between it and your stomach except a few thin layers of skin.[17]

¶ Why do I get motion sickness?

It's enough to make you spew! Although many theories exist, medical experts admit they do not fully understand what causes it. It is particularly puzzling as to why some individuals seem to be more 'motion sickness prone'. Nausea, paleness, vomiting, sweating and dizziness are all signs of motion sickness. About 90 per cent of us suffer from motion sickness at least once in our lives.[18]

Dr Mohamed Hamid, a vesitibular disorders specialist at the Cleveland Clinic Foundation in Ohio, claims that motion sickness probably originates from one of two sources. It relates either to problems in the inner ear (which controls balance) or to problems in the central nervous system (which transmits electrical impulses to the brain concerning body and head movements). According to this theory, fluid 'sloshes' through the ear's three semicircular canals simultaneously. This results in contradictory nerve impulses reaching the brain causing disequilibrium. Anxiety, stress and fatigue are contributory.[19]

Dr Cecil W.J. Hart, former chairman of the American Academy of Otorhinolaryngology, is quoted in *Australian DR Weekly* as saying that 'many people suffer from motion sickness because, when travelling, familiar cues used for orientation are upset'. For example, 'when a child is reading while riding in a car – the inner ears detect the motion of travelling, but the eyes see only the pages of print'. Dr Hart recommends that strong odours and greasy or spicy foods be avoided.[20]

Dr Kenneth Koch, a gastroenterologist at the Hershey Medical Center in Hershey, Pennsylvania, argues that there may be a motion sickness gene which makes some people more prone to the sickness than others. This is based on recent analyses of the perceptual, neurological and hormonal components of this loathsome gastric disturbance.[21]

Dr Noel Cohen, an ear, nose and throat specialist at New York University, adds that people subject to motion sickness should breathe fresh air while avoiding alcohol and caffeine. For severe cases, he recommends medication taken orally or by transdermal patch.

Dr Harold Silverman, a New York clinical pharmacologist and author of several health books, recommends the following:

|| In an aeroplane, try to get a seat over the wings.
|| On a ship, try to stay on deck as close to the middle of the ship as possible and don't focus on the motion of the waves.
|| In a train, bus, or car, try to sit facing forward and focus your gaze straight ahead.
|| Avoid reading during the trip.
|| Avoid heavy meals and alcohol.

|| If you take motion sickness medication, do so at least 30 minutes before you travel and every four to six hours during travel.[22]

Nevertheless, Dr Kenneth Dardick, an inner-ear specialist at the University of Connecticutt School of Public Health, warns that no medication has yet been developed to cure motion sickness. The best you can do is 'ride it out' as 'tolerances vary greatly from one person to another'.[23]

¶ Why don't people who take nitroglycerin for heart conditions explode?

Evacuate! Evacuate! There are bombs bursting all over the cardiac ward!

We all know that nitroglycerin is a highly explosive compound. It's a volatile chemical cocktail combining carbon, hydrogen, nitrogen and oxygen. 'Nitro' taken in pill form helps heart patients by acting directly on the wall of the blood vessels. But according to Dr Thomas Robertson, chief of the cardiac diseases branch of the National Heart, Lung and Blood Institute of the US National Institutes of Health, the amount and concentration of 'nitro' in heart medications is too small to cause any possible danger of a patient exploding. This is so even if the patient overdosed a little and jumped up and down.

Dr Robertson adds that 'nitro' is 'diluted with filler in the tablet' such that by the time 'it is absorbed in the body it is in minute concentrations. It dilates the vessels, which both increases the blood supply to the heart and reduces the work of the heart by reducing blood pressure.'[24]

¶ What causes different blood groups?

Most of us know that there are four major human blood types: A, B, AB and O. Each is divided into RH+ and RH– which makes a total of eight major blood groups in all. This is called blood polymorphism. Despite research, science does not know why we have only eight and not eighty, 800, 8,000 or even only one.

For survival as a species, it would seem to make little sense to have more than one blood group. This is because there seems to be obvious disadvantages without any advantages. For example, the best known disadvantage arises when RH– mothers have children by RH+ fathers. The mother–baby incompatibility may lead to an infant's death. It would seem that evolutionary pressure should have long since eliminated this RH incompatibility problem. But this has not happened – and it baffles science.

Disadvantages are often compensated for by advantages. For example, the haemoglobin polymorphism that gives rise to sickle-cell anaemia in Central African populations also helps to protect those populations from malaria. The trade-off is referred to as 'balanced polymorphism'.

Unfortunately, science has not been able to establish that all blood polymorphisms are balanced. In fact, we know little about nature's blood-balancing act. But we can observe that some bloods have some disadvantages over others. For example, according to research by Dr Corinne Wood, people with Type o blood, compared to those with either A, B, or AB, are more prone to typhoid fever, virus diseases (especially polio), bleeding, auto-immune diseases, and gastric ulcers. Also, malaria-carrying mosquitos prefer to take blood from Type o individuals, but why this is so remains unclear.[25]

According to research by Dr G. Jorgensen, people with Type A blood, compared to those with either o, B or AB, are more prone to malaria, cancer, smallpox, diabetes millitus, cardiac infarction, pernicious anaemia, rheumatic diseases and nephrolithiasis (a kidney problem involving the build-up of salts).[26]

But we can only go so far with such comparisons. In the 1980s, Drs J.A. Beardmore and F. Karimi-Booshehri published an article in *Nature* reporting that Type A blood was significantly more common among those in higher socio-economic groups in the UK compared to those with Type o blood. A torrent of criticism followed that has not let up since.[27]

The existence of blood polymorphism is a fact, but its meaning and biological significance remain mysteries.[28]

¶ Can a person have more than one blood type?

This raises the subject of the fascinating phenomenon of the blood chimeras.

We were all taught in high school biology that each of us is of one blood type: A, B, AB, or O – and only one type. Moreover, we were taught that it is impossible for a person to be of two blood types and that the body would reject any incompatible blood type taken in transfusion. But this does not account for blood chimeras.

A blood chimera is a human with two different blood types plus the tissues for manufacturing new blood cells for both types. Blood chimeras occur in animals and in humans – but very rarely. All known human blood chimeras are twins. It is believed that somehow, while still in the womb, three very odd things happen. Blood is shared between the twin foetuses, blood-manufacturing tissues are exchanged between them, and the immune system's normal rejection response to foreign blood is suppressed.

If your blood is Type A, your blood carries a protein called antigen A and another protein called antibody B. An antigen is a substance that stimulates the body to produce an antibody. If your blood is Type B, your blood carries antigen B and antibody A. If your blood is Type AB, your blood carries both antigens, but neither of the antibodies.

10 · **Endings**

Death happens to us all.
But there's more to death than simply to, as
Shakespeare wrote, 'shuffle off this mortal coil'.

¶ Can you die of laughter?

For the same reason, you can die of laughing. The AFP reported one case wherein a Danish doctor literally died laughing when a fit of laughter brought on a fatal heart attack at a cinema. The doctor was watching *A Fish Called Wanda*. It was believed that his heartbeat accelerated from a normal rate of 60 beats per minute to between 250 and 500.[1]

¶ Does your hair continue to grow after you die?

The conventional wisdom is that hair only seems to grow. As with supposed fingernail growth after death, hair looks as if it still grows because the body dries and shrivels up after death. The skin contracts and more of the hair is revealed.

However, there are rare cases of so-called hair 'revelation' after death far beyond that which can be attributed to mere skin retraction. For example, in their classic *Anomalies and Curiosities of Medicine* (1897), George Gould and Walter Pyle write that:

> Aristotle discusses postmortem growth of the hair, and Garmanus cites an instance in which the beard and hair were cut several times from the cadaver. We occasionally see evidence of this in the dissecting rooms. Caldwell mentions a body buried four years, the hair from which protruded at the points where the joints of the coffin had given away. The hair of the head measured 18 inches, that of the beard eight inches, and that of the breast from four to six inches...[2]

There are other such cases, most of which are a century old or more. How much credibility we can give to such accounts is up for debate.

¶ Does your hair change colour after you die?

Hair colour can change in response to the immediate chemical environment. That is why hair dyeing is so easy and so common. Sometimes hair dyeing is unintentional. For example, workers in cobalt mines sometimes have their hair turn blue and workers in copper mines have theirs turn green. But hair dyeing after dying is a horse of a different colour.

It seems that it is possible although not very common. A case originally reported in *Popular Science Monthly* in 1885 is a possible instance of hair colour changing after death. A doctor by the name of Hauptmann described the case of a body exhumed after being buried for 20 years. The hair had changed from dark brown to red. Brown is close to red on the colour spectrum, so whether or not the change was merely the fading of brown to red is worth considering.[3]

¶ Can humans 'glow in the dark' after death?

A dead body can sometimes look as if it is glowing, but really it is not. A fairly recent example of this observation comes from John Michell and Robert Rickard, who write:

> As neighbours prepared the shroud they noticed the body surrounded by a blue glow and radiating heat. The body appeared to be on fire; efforts to extinguish the luminescence failed, but eventually it faded away. On their moving the body, the sheet below it was found to be scorched.[4]

It all looks intriguing, but the explanation is rather simple. After death, the human body decomposes. Bacteria play a major part in this natural process. Some bacteria are luminous. Thus, in the rare circumstance that a decomposing human body appears to be glowing, it is probably merely the bacteria. The first time this explanation was presented was in 1838. Its two authors, Daniel and Robert Copper, conducted experiments to test various theories as to why dead humans might seem to act like glow-worms. It seems they were very curious about the 'fuel' as well as the 'fire'.[5]

Of course, if the person died by swallowing radium, then they really would glow in the tomb.

¶ Did I have a dead twin that I never knew about?

This is a very real possibility. Due to recent advances in medical technology, we now know that far more twins are conceived than are actually born. One dies in utero. This is known as the 'vanishing twin' phenomenon.

What happens to our never-born twin? How often does a twin 'vanish'?

Although first described by researchers more than a decade ago, the 'vanishing twin' has been little understood until now. Its existence is now firmly documented by ultrasound scan analyses.

Based on the same principles used in sonar during World War II to detect enemy submarines, ultrasound scanning yields a visual image on a screen showing the developing foetus from conception to birth. In fact, this image is so clear that foetal growth can be precisely measured, fingers and toes counted, foetal position shifts noted, and even urine observed in the foetal bladder. Invaluable as a medical diagnostic tool, ultrasound is like watching your unborn baby on TV (black and white only, of course, and with the volume turned off).

Interestingly, almost from the first time ultrasound was used, doctors began noticing that not all twins that were conceived were born. Ultrasounds early in the pregnancy showed twins, yet later ultrasounds revealed only one foetus. Something was happening. But what? Further observations established that whatever was happening, the disappearing twin 'vanished' sometime during the first month of pregnancy – never after the first trimester.

About a dozen 'vanishing twin' studies appeared in the medical literature beginning in 1979. The explanation for 'vanishing' attributed it to natural absorption or rejection processes of the mother's body. However, what was more intriguing about the studies was their estimations of twin disappearance rates.

Some researchers denied the existence of the phenomenon. Others estimated the rate to be as high as 78 per cent of twin conceptions.

Such a wide range in the estimations was due to several factors including differences in study technique, methodology, sample size and incorrect interpretations of either technical ultrasound artefacts or physiological conditions associated with pregnancy. These can actually mimic the presence of an additional gestational sac and thus fool the ultrasound observers.

This was further complicated by a number of confounding variables. For example, rates of twin births vary among nations. The reported incidence of twin births is highest among blacks and East Indians, followed by northern European whites. Mongolians have the lowest rate. In the UK, twins are born about once in every 80.

Other national ratios are: one in 100 in Australia, one in 86 in both the US and Italy, one in 130 in Greece, one in 150 in Japan, and one in 300 in China.

In one US study, 1,000 pregnancies were closely monitored via ultrasound and carefully analysed. It was discovered that there was a 'minimum incidence' of conceptual twinning between 3.29 per cent and 5.39 per cent – 'higher than previously believed'. Furthermore, it was found that 'vanishing' occurred in 21.2 per cent of conceived twin cases – again, higher than previously believed.[6]

'Vanishing' is very often accompanied by vaginal bleeding. It is theorised that 'vanishing' occurs when the mother's body either absorbs or rejects a foetus. In all likelihood, the foetus is absorbed or rejected due to its severe genetic defects. However, the remaining twin is not genetically defective, is left perfectly intact, and has every chance of being born completely healthy. The 'half miscarriage' of the damaged foetus explains the vaginal bleeding.[7]

¶ Can you have an 'attitude' after death?

'Attitude' refers to the position a corpse can be frozen into after sudden death. Sometimes a truly bizarre attitude can be assumed by the body at the instant of death and remain unchanged without evidence of muscle relaxation. This phenomenon is often observed on battlefields. It is quite different from normal rigor mortis.

When a body dies it is usually quite flexible and limp. Rigor mortis then sets it. The muscles of the body gradually stiffen and then gradually relax again. The cycle of rigor mortis may take 12 to 24 hours or more. The length of time depends on conditions such as cause of death, air temperature and so on.

Nevertheless, in some cases of sudden and violent death, a body will be frozen into a position that the person was in at that instant. And for some unknown reason, the muscles fail to relax and remain tense indefinitely.

This phenomenon of 'strange attitude at death' was frequently observed in nineteenth-century journalistic accounts of wars. The trench warfare of World War I provided so many cases that such gruesome descriptions were simply no longer thought worthy of telling.

Perhaps the best description of such an unusual attitude in death on the battlefield is the following:

[It] was observed by Dr Rossbach, of Wurzburg, upon the battlefield of Beaumont, near Sedan, in 1870. He found the corpse of a soldier half-sitting, half-reclining, upon the ground, and delicately holding a tin cup between his thumb and forefinger, and directing it toward a mouth that was wanting. The poor man had, while in this position, been killed by a cannonball that took off his head and all of his face except the lower jaw. The body and arms at the instant of death had suddenly taken on a rigidity that caused them to afterward remain in the position that they were in when the head was removed. Twenty-four hours had elapsed since the battle, when Dr Rossbach found the body in this state.[8]

¶ Is there such a thing as a 'near death experience'?

There's a great deal of confusion about the so-called 'near death experience' (NDE). Medical and behavioural science research is divided as to whether or not whatever is experienced is an hallucination, a product of mass hysteria, a chemical reaction in the brain due to traumatic stress, or truly 'an intermediate step between life and death'.

Improved medical technology and treatment have brought more people back from death's door than ever before. As a result, more survivors are reporting NDES. Usually they are patients who survive cardiac arrest, coma, near fatal trauma, near drowning or some other severe illness. Yet there are also cases of NDES resulting from catastrophic psychological stress not involving physical injury. Examples of this are survivors of mountain-climbing accidents and miners trapped for days after cave-ins.

Probably the first case of an NDE was reported by the ancient Greek philosopher Plato (427–347 BC). In the Republic, Plato writes of a soldier named Er who was supposedly 'killed' in battle. As the account goes, while Er lay on the battlefield, his soul took flight. Accompanied by the souls of several fellow soldiers, Er's soul experienced another world – a land where all souls were judged. He saw other souls choosing their subsequent incarnations and then drinking from the River of Forgetfulness in order to obliterate their past memories. But Er was forbidden to drink. He then blacked out. He revived into his real, conscious life just in time – his funeral pyre was being lit.[9]

¶ What is an NDE?

In 1975, Dr Raymond Moody wrote a book entitled *Life After Life* in which he coined the phrase 'near death experience'. This term was used to describe 'the common pattern of elements that people report experiencing after they have recovered' from a close brush with death. Dr Moody's book is filled with case after case of such experiences.[10]

It is now generally accepted that an NDE includes any memory a person has from the time they were unconscious – lingering between life and death – and the time they regain consciousness.

In an NDE, there are one or more of nine elements that are experienced:

|| a sense of being dead – the sudden awareness that one has had a 'fatal' accident or not survived an operation
|| peace and painlessness – a feeling that the ties that bind one to the world have been cut
|| an out-of-body experience – the sensation of peering down on one's body and perhaps seeing the doctors and nurses trying to undertake a resuscitation
|| tunnel experience – the sense of moving up or through a narrow passageway
|| 'people of light' – being met at the end of the

tunnel by others who are 'glowing'

‖ a 'Being of light' – the presence of a God-like figure or a force of some kind

‖ panoramic life review – being shown one's life by the 'Being of light'

‖ reluctance to return – the feeling of being comfortable and surrounded by the 'light', often described as 'pure love'

‖ personality transformation – a psychological change involving loss of the fear of death, greater spiritualism, a sense of 'connectedness' with the Earth and a greater zest for life.[11]

Of these, by far the most frequent element in the NDE is the 'out-of-body experience'. Perhaps 75 per cent of reported NDEs involve this. The second most frequent is the so-called 'panoramic life review'. As the term suggests, the events in one's entire life flash by in the mind in an instant. In order of diminishing frequency, the other common features of NDEs include entering a tunnel, meeting others (such as living or dead relatives), encountering 'a being of light', having a sense of the presence of 'a deity', returning to the body and experiencing what researchers call 'elements of depersonalisation' (such as an altered sense of time or a detachment from reality).

Another of the more interesting research findings

to come to light recently is the fact that there is an extraordinary similarity in the backgrounds of those who report an NDE. There is a 'consistent tendency' for them to have also reported being victims of child abuse.[12]

Many theories explaining NDEs abound and many studies are teasing out the myths from the reality. One of the more interesting findings to emerge in recent years is that there is a large cultural component in NDE. The likelihood of which elements appear differs depending on which society one was brought up in. For example, based upon the work of Dr Nsama Mumbwe of the Department of Psychology at the University of Zambia, it has been discovered that many Africans reporting an NDE see the experience as 'evil' and involving 'witchcraft'. Among Japanese who have an NDE, many report seeing long, dark rivers and beautiful flowers. These are common symbols in Japanese art. Indeed, East Indians sometimes see a 'bureaucracy' in their vision of heaven. Finally, Micronesians sometimes see heaven as a large city with skyscrapers.

Dr Kenneth Ring has identified twelve changes that often occur in a person after an NDE. These are:

|| a greater appreciation of life
|| a higher self-esteem
|| a greater compassion for others
|| a heightened sense of purpose and self-

understanding
|| a desire to learn
|| an elevated spirituality
|| a greater ecological sensitivity
|| a feeling of being more intuitive, sometimes 'psychic'
|| an increased physical sensitivity
|| a diminished tolerance to light, alcohol and drugs
|| a feeling that their brains have been 'altered' to encompass more
|| a feeling that they are now using their 'whole brain' rather than just a small part.[13]

¶ Is my chance of dying related to my astrological sign?

Astrology is pseudo science. But tell that to the millions of people who follow their star sign in newspapers every day! Or tell it to the newspaper editor who neglects to include the astrology column!

Is it possible to be born under a bad astrological sign? Do people born under one sign live longer than people born under some other sign? Generally speaking there is no such relationship. However, believe it or not, two US behavioural scientists have uncovered evidence that an astrological sign can have a bearing on your death in one way. They've shown that one's astrological sign can predict one's inclination towards considering suicide (suicide ideation).

Specifically, it's the sign of Pisces that seems to be the naughty one. Statistically, according to the two researchers, Pisces is 'significantly associated with suicide ideation'.

Dr Steven Stack from the Department of Sociology at Auburn University in Alabama and Dr David Lester from the Department of Psychology at Stockton State College in New Jersey published their findings in an article under the intriguing title of 'Born Under a Bad Sign?: Astrological Sign and Suicide Ideation'. It generated little initial attention, but the implications of their findings refuse to go away.

Traditional scientists, those who unswervingly practise the scientific method, demand hard empirical evidence at all times and scoff at the mere suggestion that astrology could have any possible predictive value for humans, have been forced to confront the findings of Drs Stack and Lester and perhaps even re-consider long-held positions. After all, part of being a scientist is to keep an open mind – open to new evidence.

'Suicide ideation' refers to the tendency to think positively about suicide and to seriously consider attempting suicide. This is regardless of whether or not the act is carried out. Suicide ideation is often the first indication of a subsequent suicide attempt. As such, it has been the object of much behavioural science attention.

Previous international research on suicide ideation has revealed some interesting findings. For example, approval of suicide behaviour is more pronounced among Catholics than Protestants, among young people more than the middle-aged or the elderly, and among people who, when young, had negative relations with their parents.

Until the Stack–Lester study, astrology was never investigated to see whether any links existed between it and suicide.

Applying sophisticated statistical techniques to 7,508

subjects drawn from the General Social Survey of the Roper Public Opinion Research Center in Storrs, Connecticut, the two researchers found that 'while astrological sun sign on the whole was not predictive of suicide attitudes ... the sun sign of Pisces is associated with suicide attitudes'. On four indicators of suicide ideation, 'respondents with the Pisces sign were more approving of suicide than the rest of the population'.[14]

Drs Stack and Lester suggest a theory to explain their findings. In both the Greek and Indian interpretations of astrology, people born under the Pisces sign 'will have lives characterised by various kinds of losses'. They add, 'persons who believe that they are more or less predestined to a life characterised by loss may, indeed, feel more depressed and hopeless. If such were the case, it would be anticipated that persons in such a depressed mood would be more approving of suicide.' The two researchers conclude that, 'whether this reflects astrological phenomena per se or socialisation factors wherein the reality of a bad sign becomes a self-fulfilling prophecy is beyond the scope of the present analysis'.

There is another possible interpretation. It could be that Drs Stack and Lester's Pisces finding is merely a statistical quirk, an anomaly, a chance, random finding without causal significance or meaning. Such quirks are common in behavioural science research.

It has been said that if you have enough chimpanzees,

enough paints, brushes and canvases and enough time, one of the chimpanzees will eventually paint the Mona Lisa.

But don't hold your breath.[15]

¶ Can humans burst into flames for no apparent reason?

This is called 'spontaneous human combustion' (SHC). Spontaneous human combustion refers to unexplained deaths or injuries by fire where the source of the fire is apparently unknown. Spontaneous human combustion has become a favourite subject of many devotees of so-called paranormal phenomena.

In fact, science tells us there is no such thing as SHC and that all such supposed cases have clear explanations.

The first supposed SHC case in the medical literature was reported by a Dr John Overton and appears in an 1836 article in the *Transactions of the Medical Society of Tennessee*. The article concerns the burn injuries of seemingly unknown origin suffered by a mathematics professor in 1835.

In *Bleak House* (1852) by Charles Dickens, the character Krook meets a macabre death by SHC. Dickens was roundly criticised in his day for (dare we say) fanning the flames of public fears concerning this phenomenon – wholly unproven then as now.

Periodically a case of alleged SHC is reported in the medical literature and more often in the public press. For example, a 1990 report from Beijing began with 'Chinese doctors are alarmed at a new medical mystery – a young boy whose body can ignite spontaneously in the most sensitive places, burning through clothing, an official newspaper reported.' It went on to add that the boy's armpits, right hand and 'private parts' were continually being burned.[16]

Perhaps the most famous case of this sort occurred in 1985. It is described at some length in *Death By Supernatural Causes* by Jenny Randles and Peter Hough.[17] In this case, after attending a cooking class, a teenage girl burst into flames while descending a stairway. She received only superficial burns to her back, but died in hospital two weeks later of complications related to infection.

Far from being an unexplained death by SHC, the probable explanation is far more mundane. Dr Bernard Knight, Professor of Forensic Pathology at the University of Wales College of Medicine, writes in *New Scientist* of this case and dismisses all such cases of alleged SHC .[18]

According to Dr Knight, 'The interest is not in that tragedy, which seems a clear instance of a gas jet igniting her clothing – but in the limitless need for people to believe in "spontaneous human combustion", a fantasy stimulated by Charles Dickens in *Bleak House*. Denied by pathologists for years, spontaneous combustion is a mass delusion that refuses to go away.'

Dr Knight adds, 'certainly, dead bodies can catch fire and burn almost to nothing – but there must be a good source of ignition and a good draught, so that the body fat acts as fuel for the wick of clothing. There is nothing "spontaneous" about this and the girl whose clothing caught fire at the back must have been too near the gas rings, albeit some minutes before, leading to minimal smouldering before flames developed.'

Hardly a paranormal phenomenon, the notion of SHC is one that science shoots down in flames.[19]

¶ Is my chance of dying related to the phase of the moon?

The power of the full moon has been blamed for bringing about insanity, accidents, suicide and murder. But is there a scientific foundation for the power of the full moon to significantly affect the way we behave?

The belief that humans are somehow affected by the full moon is one of our civilisation's oldest notions about the causes of behaviour. Both Pliny the Elder (23–79 AD) and Plutarch (46–120 AD) wrote of the widespread nature of this belief. Anthropologists report it as an extremely common notion throughout the belief systems of many non-Western societies. Our traditional folklore is filled with such references, as is our popular culture today.

Staff at psychiatric hospitals around the world have for many years claimed that patients are usually more difficult to manage during nights with a full moon. The term 'lunatic' is sustained in our language by the persistence of such reports.

We know that the full moon affects tides, plant growth and many other physical and biological processes. Women's menstruation, possibly hair and nail growth, and perhaps other aspects of human biology may be influenced as well. Statistically, more babies are born during the full moon than at any other time in the lunar cycle. Yet the reasons for this remain unclear.

Research on the possible effect of the full moon on erratic or violent behaviour has been inconclusive over the last two decades. Possibly the most convincing evidence that lunar cycles affect violent behaviour comes from Dr Arnold Lieber of the University of Miami School of Medicine. In his 1978 study, Dr Lieber concluded that the occurrence of self-destructive acts is positively correlated with the full moon. It was then theorised that this must be due to some inherent 'biological rhythm of human aggression'.[20]

Other studies hint at possible relationships but fail to establish statistically significant correlations.

Still other studies have completely rejected the full moon as an influence on erratic or violent behaviour. And this has been the trend in the research for about fifteen years. In one study, it was statistically demonstrated that the full moon has not affected the US homicide or suicide rates.[21] In another study, it was shown that the phases of the moon did not alter the rate of disruptive behaviours of inmates at a US psychiatric institution.[22]

In an unusual twist, still another study found that the number of patients with violent injuries actually fell during the full moon. This goes 180 degrees against the commonly-held view. In this study, Drs Wendy Coates, Dietrich Jehle, and Eric Cottington from the Allegheny General Hospital in Pittsburgh reviewed all admissions for trauma over a period of one year at their hospital.

They found that of the 199 stabbing and shooting victims admitted, there were eight patients admitted on full moon days for every 10 admitted on non-full moon days. This led them to write, 'we conclude that the belief in the deleterious effects of the full moon on major trauma is statistically unfounded'.[23]

¶ Is my chance of dying related to the activities of the sun?

Science knows of no direct cause and effect relationship between human mortality and the activities of the sun. Early attempts to find such correlations have been nullified by later research.

In 1972, two Russian scientists alleged that they had evidence linking solar activity with the behaviour of various living organisms. However, this evidence was not made available outside the then USSR. They even proposed a new branch of science to study the solar-behaviour relationship. But that idea was quickly hit squarely on the head only a few years later.

In 1976, Drs B.J. Lipa and P.A. Sturrock of the Institute for Plasma Research at Stanford University along with Dr F. Rogot of the National Heart and Lung Institute at the US National Institutes of Health published impressive results that solar activities have no bearing on human mortality. They examined US death rates and found no link with the sun whatsoever. They wrote, 'In summary, our study does not support the findings of Gnevyshev and Novikova, nor their proposal for a new branch of science – heliobiology.'[24]

¶ Why do people 'die in their sleep'?

There should be no real mystery in this. If people spend roughly eight hours per day sleeping and if they die of 'natural causes', then they have a one in three chance of dying during sleeping. But beyond this, there exists a strange mystery of sleep-related death that baffles medical science and defies explanation. It is called Sudden and Unexplained Death in Sleep Syndrome (SUDS).

SUDS occurs among adults, particularly among Asian adult males. No-one knows why it is more common among men, nor why Asians are the most susceptible. The US Centers for Disease Control in Atlanta labelled SUDS as the leading cause of death among young Southeast Asian men who came to the US as refugees in the early 1980s. SUDS has even been referred to as the SIDS of adults. SIDS refers to Sudden Infant Death Syndrome.

The first report of SUDS in the medical literature appeared in 1917 in the Philippines where it was known as bangungut. A 1959 report from Japan referred to the syndrome as pokkuri. It has been reported in Laos, Vietnam, Singapore and elsewhere. It has been known by various names, but it is the same strange, unexplained phenomenon.

The victims are apparently in good health just prior to

their deaths. Their tragic, sudden demise comes as a great shock to everyone. The family is often left destitute because it is often the male breadwinner who dies so suddenly.

The accounts of witnesses claim that the victim is apparently sleeping normally when all of a sudden they begin to moan, groan, snore strangely, gasp and finally choke. Technically, these are termed 'agonal signs'. Technically also, most SUDS victims die of 'ventricular fibrillation' sometimes within minutes of the agonal signs first appearing. 'Ventricular' refers to the small, lower chambers of the heart, while 'fibrillation' is a local, involuntary contraction of a muscle, invisible under the skin.

Some witnesses have tried to awaken the obviously distressed victim-to-be. But when this occurs, it is invariably unsuccessful – and the person dies just the same. When an autopsy is performed in a SUDS case, it fails to reveal any significant pathological conditions or evidence of accidental poisoning, allergy or foul play.

In 1992, seven researchers described their survey of SUDS deaths in northeastern Thailand over a two-year period. They point out that the typical pattern of SUDS in their survey is as follows: the victim dies within 24 hours of the onset of the agonal signs, is usually between 20 and 49 years of age, has 'no history of severe illness, was in good health during the previous year and was able to work during the 24-hour period before death'.[25]

Moreover, the researchers add that 'the deaths were witnessed in 63 per cent of cases and the other victims were found dead in a sleeping or resting position. In witnessed cases, 94 per cent of deaths occurred within 60 minutes of onset of acute signs or symptoms. All of the reported SUDS victims were men...'.

In addition, the victims were of normal body weight. Rates of smoking, drug usage, alcohol consumption and other possible risk factors were also normal.

Interestingly, a family history of SUDS was reported in 40.3 per cent of cases. In fact, 18.6 per cent of victims had brothers who also had died suddenly – but none had sisters who had done so.

SUDS seems to be seasonal. In Thailand at least, it is most common during March–May and least common during September–November.

The researchers note that SUDS is becoming recognised 'as a potentially important public health problem' in Thailand. It kills about one in 3000 men between the ages of 20 and 49 and ranks as a killer in this group behind accidents, poisonings, violence and heart disease.

Not surprisingly, superstition takes over at the village level in the absence of a reasonable explanation for SUDS. The researchers note that the people of rural, north-eastern Thailand refer to SUDS as laitai ('death in sleep'). The local explanation for laitai that has emerged is that a 'widow ghost' goes looking for the spirits of young men. When she finds one, she waits for the young man to sleep and then steals his spirit – causing sudden death.

The researchers note that 'fear of laitai and the "widow ghost" is now widespread in northeastern Thailand and rituals have emerged that involve disguising sleeping men with women's cosmetics, fingernail paint and bed clothes'.

One scientific explanation is that a combination of physical and psychological stressors may somehow be causing SUDS. For example, psychological factors causing related heart problems was put forward in one 1978 study. But others regard this view as highly speculative.[26]

'Widow ghost' or whatever, SUDS remains a mystery.[27]

¶ What is a zombie?

Zombies, the so-called 'walking dead', are an integral part of life in the Caribbean nation of Haiti. Among voodoo believers, zombies are reputedly corpses which have been brought back to life through black magic. The corpse is allegedly revived after the power of the snake or python deity enters the body. The corpse, now a zombie, is still regarded as a human being, but one without speech, will or judgement. Nevertheless, the zombie is capable of automatic movement such as walking and following simple commands. It is a trance-like, sleep-like state. Indeed, 'walking dead' is very descriptive.

The origin of the belief in zombies is lost in antiquity. However, it is almost certainly derived from the traditional religious and folklore traditions of West Africa – where Haitians came to the New World as slaves.

Although educated Haitians and Westerners regard zombies as imaginary beings, perhaps on a par with the werewolves of Romania or the leprechauns of Ireland, most Haitians regard zombies as very real. In fact, even today few Haitians dare to violate the voodoo taboo which forbids even mentioning the subject. Indeed, the making of zombies – 'zombification' – is a term recognised in Haitian law.

Dr Lamarque Douyon, a Haitian psychiatrist in Port-au-Prince, has studied the phenomenon of zombies for nearly two decades. Where other Haitian physicians refuse to attend so-called zombies who also happen to need medical treatment (no doubt in fear of the voodoo taboo against doing this too!), Dr Douyon has treated many such people. Placing science, understanding and devotion to humanity ahead of fear and superstition, he believes he has solved the mysteries surrounding zombies.

It seems that making a zombie involves simple biochemistry and native folk pharmacology. Dr Douyon claims that creating a zombie from a normal person is merely a question of drugging them up without their knowledge. In fact, zombification is generally used as a punishment imposed by voodoo courts. Haitians, so it seems, as often as not turn to voodoo courts for justice since the official Haitian legal system is seen as corrupt and serving only the rich. Most Haitians prefer to resort to voodoo courts for the resolution of a variety of disputes ranging from property matters to personal grievances. Thus, a person may be made into a zombie as the court's punishment in much the same way as a magistrate's court may impose a fine or period of imprisonment.

According to Dr Douyon, the transformation of a person into a zombie is accomplished by having them drink one of several kinds of poison. The first, tetrodotoxin, has the effect of putting normal human beings into a death-like trance. This poison is extracted from a ball-shaped fish, the tetrodon, and combined with severed parts of a dead toad for good measure. The second drug is made from a normally inedible vegetable appropriately named the zombie cucumber (concombre zombi). This poison has the effect of slowing the metabolism to the point where the victim appears to be dead. There is yet another drug, still unknown to Dr Douyon, but very much known to voodoo priests he claims, which gives the already comatose-like person a more normal metabolism and physical functioning ability.

Dr Douyon stresses that when the poison is administered, the victim soon appears to be dead. The bereaved family, unaware that the victim is still alive, often proceeds with a burial. Later, after the zombie is revived, he or she returns in a disoriented, stuporous condition which gives rise to the myths of zombies 'walking dead'.

Haitian funeral and burial traditions permit zombification to persist. In fact, proper etiquette calls for family and friends to leave the cemetery before the actual burial of a body takes place. According to Dr Douyon, this allows all manner of foul play to occur. A body may be switched, abducted, revived and so on without it being discovered.[28]

¶ How do you make a dead body into a mummy?

Embalming and creating mummies originated with the Egyptians about 3,000 BC. 'Mummification' is from a Persian word meaning 'wax'. The techniques of mummification developed by trial and error. Earlier mummies tend not to be as well done as later ones.

The oldest known mummy is that of a young princess who was buried around 2,600 BC on a plateau near the Great Pyramid of Cheops at Giza. The mummy was excavated in 1989. The oldest complete mummy is that of a court musician who was buried around 2,400 BC in the Nefer tomb of Saqqara. It was unearthed in 1944.

Generally speaking, it is believed that the corpse was preserved with aromatic resins from plants of the genus Commiphora. The principal resin used was balm. This is where we get the word 'embalming'.

The earliest embalming method called for the corpse to be wrapped in cloth and buried in charcoal and sand in an area free of humidity. But by the beginning of the New Kingdom, around 1,570 BC, embalming was such an art form that ancient Egyptian morticians competed against each other. They tried to turn out the most realistic visage of life possible, often through painting the face and other body adornments.

The embalming of ancient Egyptian royalty involved major surgery. The heart, lungs and intestines were excised, washed in palm wine and sealed in urns filled with alcohol and herbs.

Interestingly, the brain was the one organ that was discarded. It was believed to be merely a functionless mass – regarded in the same way as we regard fat.

Myrrh, other fragrant resins and oils, as well as specially prepared perfumes, were poured into the eviscerated body cavity and the incision stitched closed. The body was packed in saltpetre, an astringent, for two months, removed, soaked in wine, swathed snugly in cotton bandages, dipped in a porridge-like, anti-bacterial paste, lowered into an ornate coffin and housed in a sepulchre.

The ancient Egyptians were such excellent surgeons that brain surgery was not beyond their scope. They learned such skills in order to preserve the dead as much as to cure the living. Egyptian mummification and embalming techniques were lost by the beginning of Christianity.

No human society has ever achieved immortality. The Egyptians certainly tried.[29]

References

1

Beginnings

1 Kimbel, W., Johanson, D. & Rak, Y., 'The First Skull and Other New Discoveries of Australopithecus Afarensis at Hadar, Ethiopia', *Nature*, 1994, 368: 449-451.

2 White, T., Suwa, G. & Asfaw, B., 'Australopithecus Ramidus: A New Species of Early Hominid from Aramis, Ethiopia', *Nature*, 1994, 371: 306-312.

3 Chamberlain, D., 'The Sentient Prenate: What Every Parent Should Know', *Pre- and Perinatal Psychology Journal*, 1994, 9: 9-31, pp 12-13.

4 Birnholz, J., Stephens, J. & Faria, M., 'Fetal Movement Patterns: A Possible Means of Defining Neurologic Development Milestones in Utero', *American Journal of Roentology*, 1978, 130: 537-540.

5 Liley, A., 'The Foetus as a Personality', *Australian and New Zealand Journal of Psychiatry*, 1972, 6: 99-105.

6 Birnholz, J., 'The Development of Human Fetal Eye Movement Patterns', *Science*, 1981, 213: 679-681.

7 Anand, K. & Hickey, P., 'Pain and its Effects in the Human Neonate and Fetus', *New England Journal of Medicine*, 1987, 317: 1321-1329.

8 Hooker, D., *The Prenatal Origins of Behaviour*, New York, Hafner, 1969.

9 Juan, S., 'For Baby, A Touch of Love Will Do Nicely', *The Sydney Morning Herald*, 14 December 1989, p 17.

10 Morse, M., 'Life Inside the Womb', *Parents*, December 1994, pp 74-76, p 76, quoted in Juan, S., 'How We've Been Thick-Skinned About Touching', *The Sydney Morning Herald*, 9 November 1989, p. 13.

11 *ibid.*

12 Peleg, D. & Goldman, J., 'Fetal Heart Rate Acceleration in Response to Light Stimulation as a Clinical Measure of Foetal Wellbeing: A Preliminary Report', *Journal of Peri-Natal Medicine*, 1980, 8: 38-41.

13 Juan, S., 'Baby's Eye-View Described', *Eye Care Australia*, October 1989, p 8.

14 Shahidullah, S. & Hepper, P., 'Hearing in the Fetus: Prenatal Detection

of Deafness', *International Journal of Prenatal & Perinatal Studies*, 1992, 4: 3–4: 235–240.

15 Chamberlain, D., *op. cit.*

16 Wertheimer, M., 'Psychomotor Coordination of Auditory and Visual Space at Birth', *Science*, 1961, 134: 1692–1693.

17 Morse, M., *op. cit.*

18 Disher, D., 'The Reactions of Newborn Infants to Chemical Stimuli Administered Nasally', in Dockeray, F., ed., *Studies in Infant Behavior*, Columbus, Ohio State University Press, 1934, pp 1–52.

19 Steiner, J., 'Human Facial Expression in Response to Taste and Smell Stimulation', in Reese, H. & Lipsitt, L., eds., *Advances in Child Development and Behavior*, New York, Academic Press, 1979, pp 257–295.

20 Maurer, D. & Maurer, C., *The World of the Newborn*, New York, Basic Books, 1988, pp 86–87, quoted in Juan, S., 'How Does a Baby Smell? Pretty Well, By All Reports', *The Sydney Morning Herald*, 16 November 1989, p 16.

21 Tatzer, E., Schubert, M., Timischl, W. & Simbruner, G., 'Discrimination of Taste and Preference for Sweet in Premature Babies', *Early Human Development*, 1985, 12: 23–30.

22 Morse, M., *op. cit.*

23 DeSnoo, K., 'Das Trinkende Kind Im Uterus', *Monatsschrift Fur Geburtshilfe & Gynaekologie*, 1937, 105: 88–97.

24 Freed, F., 'Report of a Case of Vagitus Uterinus', *American Journal of Obstetrics & Gynecology*, 1927, 14: 87–89.

25 Juan, S., 'About Face', *24 Hours*, January 1991, p 20.

26 Hepper, P., Shaidullah, S. & White, R., 'Origins of Fetal Handedness', *Nature*, 1990, 347: 431.

27 Luria, A., *The Mind of a Mnemonist*, New York, Avon Books, 1969.

28 Hoffmann, R., 'Developmental Changes in Human Infant Visual-Evoked Potentials to Patterned Stimuli Recorded at Different Scalp Locations', *Child Development*, 1978, 49: 110–118.

29 Maurer & Maurer, *op. cit.*, pp 65–66.

30 Lewkowicz, D. & Turkewitz, G., 'Cross-Modal Equivalence in Early Infancy: Auditory-Visual Intensity Matching', *Developmental Psychology*, 1980, 16: 597–607.

31 Maurer & Maurer, *op. cit.*, pp 66–67.

32 Cytowic, R., *Synesthesia: A Union of the Senses*, New York, Springer-

Verlag, 1989, and *The Man Who Tasted Shapes*, New York, Jeremy
Tarcher/Putnam, 1993.

33 Baron-Cohen, S., Wyke, M. & Binnie, C., 'Hearing Words and Seeing
Colours: An Experimental Investigation of a Case of Synaesthesia',
Perception, 1987, 16: 761–767.

34 Motluk, A., 'The Sweet Smell of Purple', *New Scientist*, 13 August 1994,
pp 32–37, quoted in Juan, S., 'When Commonsense is Nonsense',
The Sydney Morning Herald, 5 October 1989, p 13.

35 Hansen, J., '"Blindsight" — Seeing Without Realizing That You Can See',
Science Digest, January 1980, p 14.

36 Horgan, J., 'Can Science Explain Consciousness?', *Scientific American*,
July 1994, pp 88–94, p 91.

37 Caudill, M., *In Our Own Image: Building An Artificial Person*, Melbourne,
Oxford University Press, 1992.

38 Martino, J., 'Robotic Luggage Carrier', *The Futurist*, July/August 1992,
p 6.

39 Tarlow, P. & Muehsam, M., 'Wide Horizons: Travel and Tourism in the
Coming Decades', *The Futurist*, November/December 1992, pp 28–32.

40 Martino, J., 'Artificial Muscle', *The Futurist*, November/December 1992,
p 8.

41 Juan, S., 'Intelligent Androids: The Next Generation',
The Sydney Morning Herald, 10 December 1992, p 16.

42 Rosenfeld, A., 'The Medical Story of the Century', *Longevity*, May 1992,
pp 42–53, p 53.

43 ibid.

44 Juan, S., 'Soon the Old May Be Able to Grow Younger',
The Sydney Morning Herald, 21 May 1992, p 12.

45 Kahn, C., 'Long-Life Forecast: Health Predictions You Can Use Now',
Longevity, March 1995, p 14.

46 Perls, T., 'The Oldest Old', *Scientific American*, January 1995,
pp 70–75, p 70.

2

The Brain

1 Sturge-Weber Foundation, 'Sturge-Weber Syndrome: An Overview for Physicians and Families' (video), Aurora, Colorado, Sturge-Weber Foundation, 1994.

2 McCutcheon, M., *The Compass in Your Nose*, Melbourne, Schwartz & Wilkinson, 1989, pp 64–65.

3 United Press International, 'Study Finds More Southpaws among Gays and Lesbians', *The San Francisco Chronicle*, 26 July 1990, p B3.

4 Schmeck, H., 'Q & A', *The New York Times*, 7 June 1988, p B7.

5 Juan, S., 'Get It Right or Die Trying', *The Sydney Morning Herald*, 4 August 1993, p 14.

6 McMonnies, C., 'Left-Right Discrimination in Adults', *Clinical and Experimental Optometry*, 1990, 73: 155–158.

7 McMonnies, C., 'Visual-Spatial Discrimination and Mirror Image Letter Reversals in Reading', *Journal of the American Optometric Association*, 1992, 63: 698–704.

8 Schmeck, H., 'Q & A', *The New York Times*, 23 February 1988, p Y19.

9 Browne, M., '"Handedness" Seen in Nature, Long Before Hands', *The New York Times*, 15 June 1993, pp B5, B7.

10 Halpern, D. & Coren, S., 'Longevity and Handedness', *Nature*, 1988, 333: 213.

11 Anderson, M., 'Lateral Preference and Longevity', *Nature*, 1989, 341: 112, and United Press International, 'Lefties Live Longer, Newest Study Says', *The San Francisco Chronicle*, 14 September 1989, p B3.

12 Stroh, M., 'It's Risky to be a Lefty', *Science News*, 23 May 1992, p 351.

13 Bristow, D., 'Left in Doubt', *New Scientist*, 28 May 1994, p 65.

14 Coffin, G., 'Asymmetry of the Human Head: Clinical Observations', *Clinical Pediatrics*, 1987, 25: 230–232.

15 Juan, S., '"Water" on the Brain: Facts and Myths', *The Sydney Morning Herald*, 28 November 1991, p 17.

16 Juan, S., 'Sharing Memories May Be as Simple as Sharing a Needle', *The Sydney Morning Herald*, 9 February 1994, p 13.

17 Stock, G., Metaman: The Merging of Humans and Machines into a Global Superorganism, New York, Simon & Schuster, 1993.

18 Moravec, H., *Mind Children*, Cambridge, Massachusetts, Harvard
    University Press, 1988, pp 22-23.

19 Juan, S., 'And You Thought Computer Crashes Were a Headache
    Now ...',*The Sydney Morning Herald*, 23 March 1994, p 13.

3
The Head

1   Brennan, T., Funk, S. & Frothingham, T., 'Disproportionate Intra-Uterine
    Head Growth and Developmental Outcome', *Developmental Medicine and
    Child Neurology*, 1985, 27: 746-750.

2   Elliman, A.M., Bryan, E., Elliman, A.D. & Starte, D., 'Narrow Heads of
    Preterm Infants – Do They Matter?', *Developmental Medicine and Child
    Neurology*, 1986, 28: 745-748.

3   Edell, D., 'Hole in the Head', *The People's Medical Journal*,
    August 1986, p 8.

4   Smith, W., *Bore Hole*, New York, Tribune Press, 1964.

5   Juan, S., 'Human Skull Theories That Could Turn Heads',
    Times On Sunday, 5 April 1987, p 31.

6   Fleming, C., 'If We Can Keep A Severed Head Alive...',
    British Medical Journal, 1988, 297: 1048.

7   Fleming, C., *If We Can Keep A Severed Head Alive*, St. Louis,
    Polinym Press, 1988.

8   White, R., Wolin, L., Massopust, L., Taslitz, N. & Verdura, J., 'Cephalic
    Exchange Transplantation in the Monkey', *Surgery*, 1971, 70: 135-139.

9   Hamblin, T., 'Nipping in the Bud', *British Medical Journal*, 1988, 297: 629.

10 Juan, S., 'There's Life in the Old Head', *The Sydney Morning Herald*,
    26 April 1990, p 15.

11 Adebonojo, F., 'Infant Head Shaping', *Journal of the American Medical
    Association*, 1991, 265: 1179.

12 Epstein, F., Hochwald, G. & Ransohoff, J., 'Neonatal Hydrocephalus
    Treated By Compressive Head Wrapping', *The Lancet*, 1973, I: 634-636.

13 Gould, S., *The Mismeasure of Man*, London, W.W. Norton, 1981.

14 Juan, S., 'Had a Facelift? Then Try a Reshaped Skull',
    The Sydney Morning Herald, 5 November 1992, p 12.

15 Sugar, O., 'Head Shrinking', *The Journal of the American Medical*

Association, 1971, 216: 1: 117–120.

16   Karsten, R., 'The Head-Hunters of Western Amazonas', Societas
    Scientiarum Fennica, Commentationes Humanarum Litterarum, 1935,
    7: 1–588, and Stirling, M., 'Historical and Ethnological Material on the
    Jivaro Indians', *Bulletins of the Bureau of American Ethnology*, 1938,
    117: 1–148.

17   Leavesley, J., 'Headshrinking as an Art Form', *Australian Doctor*,
    9 September 1994, p 81.

18   ibid.

19   ibid.

20   ibid.

21   Gould, S., *Ever Since Darwin*, New York, W.W. Norton, 1977, p 63.

4
The Eyes

1   Ferrer, H., 'Voluntary Propulsion of Both Eyeballs', *American Journal of
    Ophthalmology*, 1928, 11: 883.

2   Smith, J., 'Voluntary Propulsion of Both Eyeballs', *Journal of the American
    Medical Association*, 1932, 98: 398.

3   Berman, B., 'Voluntary Propulsion of the Eyeball: The Double Whammy
    Syndrome', *Archives of Internal Medicine*, 1966, 117: 648–651.

4   Borgquist, A., 'Crying', *American Journal of Psychology*, 1906, 27: 149–205.

5   Levoy, G., 'Crying It Out', *The San Jose Mercury News*, 2 November 1988,
    pp 1F; 3F.

6   ibid.

7   Carey, B., 'Vital Statistics', *Hippocrates*, September/October 1988, p 16.

8   Rymer, R., 'Why Do Women Cry More Than Men?', *Hippocrates*,
    January/February 1989, p 100.

9   Carey, B., op. cit.

10   Juan, S., 'Why a Good Cry Can Make You Feel Better', *The Sydney* Morning
    Herald, 28 October 1988, p 13.

11   Juan, S., 'How to Spot Some Signs of Aging', *The Sydney Morning Herald*,
    13 February 1989, p 15.

12   Troiano, L., 'The Wink of an Eye', *American Health*, November 1989,
    p 40.

13 Albert, B., 'What Causes Bags and Rings Under the Eyes?', *University of California, Berkeley Wellness Letter*, June 1989, p 8.

14 ibid.

15 Bower, B., 'Vision System Puts Eyesight in Blind Spots', *Science News*, 27 April 1991, p 262, and Lipkin, R., 'Focusing the Soul's Fuzzy Window', *Science News*, 11 September 1993, p 172.

16 Wright, S., 'How Carrots Can Aid Eyesight', *The San Jose Mercury News*, 13 February 1990, p C1.

17 Fackelmann, K., 'Nutrients May Prevent Blinding Disease', *Science News*, 12 November 1994, p 310.

18 Fackelmann, K., 'Nocturnal Risks For the Eyes', *Science News*, 8 May 1993, p 302.

19 ibid.

20 Franklin, D., 'Tuning the Kids In', *In Health*, December/January 1992, p 40.

21 Siebers, T., *The Mirror of Medusa*, Berkeley, University of California Press, 1983.

22 Maloney, C., (ed.), *The Evil Eye*, New York, Columbia University Press, 1976.

23 Juan, S., 'People Still Feel Deadly Eye's Power', *The Sydney Morning Herald*, 8 August 1991, p 12.

24 Juan, S., 'How Soon Can You Tell Colour?', *Eye Care Australia*, September 1989, p 12.

5
The Nose, Ears and Mouth

1 Blakeslee, S., 'Q & A', *The New York Times*, 12 January 1988, pY19.

2 Provine, R., Hamernik, H. & Curchack, B., 'Yawning: Relation to Sleeping and Stretching in Humans', *Ethology*, 1987, 76: 152–160.

3 Rensberger, B., 'This Story Is Bound to Make You Yawn', *The Washington Post*, 12 November 1992, pp D1, D6.

4 Juan, S., 'Why One Yawn Can Lead To Another', *The Sydney Morning Herald*, 14 July 1988, p 16.

5 Juan, S., 'When It's All in the Ear of the Beholder', *The Sydney Morning Herald*, 9 August 1988, p 17.

6 Blakeslee, S., 'Q & A', *The New York Times*, 8 December 1987, p Y26.

7 Davis, L., 'What's That Ringing in My Ears?', *In Health*,
   September/October 1991, p 100.

8 Margaretten-Ohring, J., 'Tinnitus: When Your Ears Ring and Ring',
   University of California, Berkeley Wellness Letter, March 1990, p 7.

9 Juan, S., 'When All That Hissing and Buzzing Becomes Too Much',
   The Sydney Morning Herald, 24 August 1988, p 17.

10 Cooper, M., *Winning With Your Voice*, Hollywood, Florida, Fell Publishers,
    1989.

11 Moody, L., 'What Your Voice Says About You',
    The Los Angeles Daily News, 13 May 1992, pp D2, D3.

12 Zamichow, N., 'The Urge To "Um"', *The Los Angeles Times*, 29 April 1992,
    p B4, and Juan, S., 'Sick of the Sound of Your Own Voice', *The Sydney
    Morning Herald*, 12 November 1992, p 12.

13 Chui, G., 'The Bulbous Red Nose', *The San Jose Mercury News*, 20 June
    1989, p 2C.

14 Goldsmith, M., 'New Topical Therapy for Acne Rosacea Offers
    Conspicuous Improvement, No Systemic Effects', *Journal of the American
    Medical Association*, 1989, 261: 2014-2015.

15 Juan, S., 'A Cure for Rudolph', *The Sydney Morning Herald*, 7 September
    1989, p 12.

16 Xenakis, A., *Why Doesn't My Funny Bone Make Me Laugh?*, New York,
    Villard Books, 1993, pp 85–86.

17 Blakeslee, S., 'Q & A', *The New York Times*, 3 October 1987, p Y16.

18 Juan, S., 'A Hard-to-Swallow Ailment', *The Sydney Morning Herald*,
    24 November 1993, p 13.

19 Ray, C., 'Q & A', *The New York Times*, 23 June 1992, p B8.

20 Hart, C., 'Light Sneeze', *New Scientist*, 25 June 1994, p 65.

21 Eccles, R., 'Light Sneeze', *New Scientist*, 21 May 1994, p 57.

6
The Skin

1 Syme, S., 'The Itch That Dares Not Speak Its Name', *University of California, Berkeley Wellness Letter*, March 1988, pp 6–7.

2 Juan, S., 'Irritating Mystery of Itching', *The Sydney Morning Herald*, 12 September 1988, p 17.

3 Halpern, D., Blake, R. & Hillenbrand, J., 'Psychoacoustics of a Chilling Sound', *Perception & Psychophysics*, 1986, 39: 77–80.

4 Juan, S., 'How a Call From the Past Can Give You the Shivers', *The Sydney Morning Herald*, 5 October 1988, p 17.

5 Schmeck, H., 'Q & A', *The New York Times*, 17 February 1987, p Y19.

6 Juan, S., 'Thin-Skinned, Lots of Exposure and Likely to Wrinkle', *The Sydney Morning Herald*, 14 November 1988, p 21.

7 Juan, S., 'Wrinkles? That's Stretching It a Bit...', *The Sydney Morning Herald*, 8 June 1994, p 13, and Blakeslee, S., 'Q & A', *The New York Times*, 23 February 1988, p Y19.

8 Brewer, S., 'Fountain of Youth For Aging Skin', *Longevity*, January 1994, p 5.

9 Xenakis, A., *Why Doesn't My Funny Bone Make Me Laugh?*, New York, Villard Books, 1993, pp 140–142.

10 Noll, R., 'Hypnotherapy For Warts in Children and Adolescents', *Journal of Developmental and Behavioral Pediatrics*, 1994, 15: 170–173, p. 170.

11 Juan, S., 'The War to End All Warts', *The Sydney Morning Herald*, 22 June 1994, p 11.

12 Fackelmann, K., 'Genital-Wart Virus Linked to Penile Cancer', *Science News*, 23 May 1992, p 342.

13 Reinisch, J., 'Genital Warts Caused by Virus', *The San Francisco Chronicle*, 23 June 1992, p B5.

14 Siegel, D. & McDaniel, S., 'The Frog Prince: Tale and Toxicology', *American Journal of Orthopsychiatry*, 1991, 61: 558–562, p 558.

15 ibid., quoted in Juan, S., 'Prince Charming Was Just a Hallucination', *The Sydney Morning Herald*, 7 May 1992, p 14.

16 Juan, S., 'When Your Hair Stands On End', *The Sydney Morning Herald*, 25 January 1989, p 13.

17 Staff of the Royal Children's Hospital Melbourne, Australia, *Paediatric Handbook*, Fourth Edition, Melbourne, Blackwell Scientific Publications, 1992, p 159.

18 Blakeslee, S., 'Q & A', *The New York Times*, 31 March 1987, p Y21, quoted in Juan, S., 'It's Just a Question of Degrees', *The Sydney Morning Herald*, 24 March 1989, p 10.

19 Wilford, J., 'Q & A', *The New York Times*, 23 August 1988, p B8.

20 Juan, S., 'Why Men Have Nipples', *The Sydney Morning Herald*, 31 December 1990, p 11.

21 Diamond, J., 'Father's Milk', *Discover*, February 1995, pp 82–87, p 82.

22 Flinn, J., 'Are You Blushing Yet?', *The San Francisco Examiner*, 28 January 1990, pp E18-E17.

23 Scott, J., 'A Blush Means, "I'm Sorry" ', *The Los Angeles Times*, 26 January 1990, pp D1-D4.

24 Juan, S., 'Blushing: Why We Do It', *The Sydney Morning Herald*, 4 February 1992, p 10.

25 Curtis, J. & Lawrence, K., 'Comes in Handy', *New Scientist*, 6 July 1994, p 65.

26 Nordlund, J., *Guidelines for the Treatment of Patients With Vitiligo*, Tyler, Texas, National Vitiligo Foundation, Inc., 1992, p 2.

27 Juan, S., 'A Black and White Issue', *The Sydney Morning Herald*, 17 February 1993, p 10.

7
The Hair and Nails

1 Barnhurst, B., 'Grey Matters', *New Scientist*, 4 February 1995, p 57.

2 Parachini, A., 'Root of Graying Remains a Mystery for Scientists', *The Los Angeles Times*, 13 October 1987, pp 1E, 3E.

3 Juan, S., 'Grey Hair: Just an Optical Illusion', *The Sydney Morning Herald*, 4 November 1988, p 17.

4 Juan, S., 'Fact That Will Make Most of Your Hair Curl', *The Sydney Morning Herald*, 22 February 1989, p 19.

5 Addison, W., 'Beardedness As a Factor in Perceived Masculinity', *Perceptual and Motor Skills*, 1989, 68: 921–922, p 921.

6 Muller, S., 'Trichotillomania: A Histopathological Study in Sixty-Six

Patients', *Journal of the American Academy of Dermatology*, 1990, 23: 56–62.

7   Christenson, G., Pyle, R. & Mitchell, J., 'Estimated Lifetime Prevalence of Trichotillomania in College Students', *Journal of Clinical Psychiatry*, 1991, 52: 415–417, p 415.

8   Rapoport, J., *The Boy Who Couldn't Stop Washing*, New York, Dutton, 1989, pp 152–153.

9   Repinski, K., 'Thicker, Youthful Hair — Another Drug to Save It?', *Longevity*, January 1992, p 20, quoted in Juan, S., 'This Can Make You Tear Your Hair Out', *The Sydney Morning Herald*, 6 February 1992, p 12.

10 Samman, P. & Fenton, D., *Nails in Disease*, Fourth Edition, St Louis, C.V. Mosby, 1986.

11 Scher, R. & Daniel, D., *Nails: Therapy, Diagnosis and Surgery*, Philadelphia, W.B. Saunders, 1990.

12 Juan, S., 'Some Theories On Nail-Biting to Chew Over', *The Sydney Morning Herald*, 28 April 1993, p 18.

13 Mauvais-Jarvis, P., Kuttenn, F. & Mowszowicz, I., *Hirsutism*, New York, Springer-Verlag, 1981.

14 Judd, S. & Carter, J., 'The Changing Face of Hirsutism', *Medical Journal of Australia*, 1992, 156: 148–149.

15 Zwicker, H. & Rittmaster, R., 'Androsterone Sulfate: Physiology and Clinical Significance in Hirsute Women', *Journal of Clinical Endocrinology and Metabolism*, 1993, 76: 112–116.

16 Juan, S., 'The Hirsute of Excellence: A Hair Question of What's Fashionable', *The Sydney Morning Herald*, 20 October 1993, p 15.

17 Deslypere, J., Praet, M. & Verdonk, G., 'An Unusual Case of the Trichobezoar: The Rapunzel Syndrome', *American Journal of Gastroenterology*, 1982, 77: 467–470.

18 Kirk, A., Bowers, B., Moylan, J. & Meyers, W., 'Toothbrush Swallowing', *Archives of Surgery*, 1988, 123: 382–384.

19 Williams, R., 'The Fascinating History of Bezoars', *Medical Journal of Australia*, 1986, 145: 613–614.

20 Juan, S., 'Why the Hair Ball's Charm Is a Hard One to Swallow', *The Sydney Morning Herald*, 28 July 1988, p 15, and Juan, S., *Only Human*, Sydney, Random House Australia, 1990, pp 150–151.

21 Brooks, A., 'Shiny, Happy Hair', *New Scientist*, 6 August 1994, p 65.

8

The Skeleton, Bones and Teeth

1 Angert, E., Clements, K. & Pace, N., 'The Largest Bacterium', *Nature*, 1993, 362: 239–241.

2 Juan, S., 'When Big Just Keeps Getting Bigger', *The Sydney Morning Herald*, 22 September 1993, p 15.

3 Samaras, T., 'Trends Toward Tallness', *The Futurist*, January/February 1995, p 29.

4 Juan, S., 'Brains But No Brawn', *The Sydney Morning Herald*, 6 October 1993, p 17.

5 Goleman, D., 'Q & A', *The New York Times*, 7 July 1987, p Y15.

6 Carey, B., 'How Are People Able to Predict Bad Weather by Pain in Their Joints?', *Hippocrates*, January/February 1989, p 100.

7 Juan, S., 'Weather Puts Arthritics Out of Joint', *The Sydney Morning Herald*, 6 July 1989, p 14.

8 Tributsch, H., *When the Snakes Awake: Animals and Earthquake Prediction*, Cambridge, Massachusetts, M.I.T. Press, 1982.

9 Raskin, D., 'Earthquakes: Do Animals Know?', *American Health*, May 1990, p 102.

10 ibid.

11 Wallace, I., Wallechinsky, D. & Wallace, A., 'Animals That Foretell Quakes', *The San Francisco Chronicle*, 23 May 1990, p B3.

12 Juan, S., 'Can Our Bodies Warn Us of Earthquakes?', *The Sydney Morning Herald*, 17 May 1990, p 17.

13 Jones, P., '"Dickens" Literary Children', *Australian Paediatric Journal*, 1972, 8: 233–245.

14 Associated Press, 'Tiny Tim Had Kidney Disease, Doctor Says', *The San Francisco Chronicle*, 15 December 1992, p 3.

15 Juan, S., 'What May Have Ailed Tiny Tim', *The Sydney Morning Herald*, 20 December 1990, p 12.

16 Joyce, C. & Stover, E., *Witnesses From the Grave: The Stories Bones Tell*, Boston, Little, Brown And Company, 1991.

17 Chui, G., 'Bones Yield Clues to Solve Mysteries of the Ages', *The San Jose Mercury News*, 5 August 1991, pp 1A–1B.

18 Juan, S., 'Secrets of Identity Easily Unlocked',

The Sydney Morning Herald, 10 October 1991, p 16.

19 Warne, G., 'Contemporary Issues in the Use of Growth Hormone', *Journal of Paediatrics and Child Health*, 1990, 26: 122-123.

20 Fackelmann, K., 'Pygmy Paradox Prompts a Short Answer', *Science News*, 8 July 1989, p 22.

21 Hallett, J., *Pygmy Kitabu*, New York, Random House, 1973.

22 Reuters, 'Last Pygmies in Danger', *Australian DR Weekly*, 15 May 1992, p 48, quoted in Juan, S., 'The Long and Short of Life As a Pygmy', *The Sydney Morning Herald*, 2 July 1992, p 16.

23 Edell, D., 'Q & A With Dr Edell', *The Edell Health Letter*, March 1993, p 8.

24 Juan, S., 'Do We Really Have a "Funny Bone"?', *The Sydney Morning Herald*, 14 April 1993, p 12.

25 Edell, D., 'Q & A With Dr Edell', *The Edell Health Letter*, March 1991, p 8.

26 Ezzell, C., 'Writer's Cramp: Literally in Your Head', *Science News*, 23 November 1991, p 333, quoted in Juan, S., 'Do We Really Have a "Funny Bone"?', op. cit.

27 Rosenthal, E., 'Q. & A', *The New York Times*, 29 December 1987, p Y17.

28 Melzack, R., 'Phantom Limbs', *Scientific American*, April 1992, pp 120-126, Feldman, S., 'Phantom Limbs', *American Journal of Psychology*, 1940, 53: 590-598, and Corliss, W., *Biological Anomalies: Humans II*, Glen Arm, Maryland: The Sourcebook Project, 1993, pp 205-207.

29 Edell, D., 'Q & A With Dr Edell', *The Edell Health Letter* December/January 1993, p 8.

30 Riggs, L., 'Men Shrink an Average of 1¼ Inches', *Bottom Line*, 1 February 1995, p 6.

31 Wales, J. & Dangerfield, P., 'Night Growth', *New Scientist*, 18 February 1995, p 65.

9
The Inside

1 Schweiger, A. & Parducci, A., 'Nocebo: The Psychologic Induction of Pain', *Pavlovian Journal of Biological Science*, 1981, 16: 140–143.
2 Morse, J. & Morse, R., 'Cultural Variation in the Inference of Pain', *Journal of Cross-Cultural Psychology*, 1988, 19: 232–242.
3 Juan, S., 'How We Can Feel Pain From Experience', *The Sydney Morning Herald*, 1 December 1988, p 16.
4 Cousins, N., *Anatomy of an Illness as Perceived by the Patient*, Toronto, Bantam Books, 1979.
5 Lang, S., 'Laughter Is the Best Defense', *American Health*, December 1988, p 42.
6 ibid., quoted in Juan, S., 'Heard the One About the Immune System?', *The Sydney Morning Herald*, 18 May 1989, p 15.
7 Lang, S., 'Laughing Matters — At Work', *American Health*, September 1988, p 46.
8 Peterson, C., Seligman, M. & Vaillant, G., 'Pessimistic Explanatory Style Is a Risk Factor for Physical Illness: A Thirty-Five-Year Longitudinal Study', *Journal of Personality and Social Psychology*, 1988, 55: 23–27.
9 Juan, S., 'Laughing All the Way to the Top', *The Sydney Morning Herald*, 22 September 1988, p 14.
10 Ziv, A., 'Teaching and Learning With Humor: Experiment and Replication', *Journal of Experimental Education*, 1988, 57: 5–15.
11 Dimmer, S., Carroll, J. & Wyatt, G., 'Uses of Humour in Psychotherapy', Psychological Reports, 1990, 66: 795–801.
12 Fry, W. & Salameh, W., (eds), *Handbook of Humour in Psychotherapy*, New York, Pergamon, 1987, quoted in Juan, S., 'Research Into Humour No Laughing Matter', *The Sydney Morning Herald*, 20 June 1991, p 14.
13 Bass, S., 'Wearing a Smile May Cheer You Up', *The New York Times*, 1 August 1989, p B1.
14 Stone, J., 'What's in Her Smile?', *American Health*, September 1990, pp 30–35.
15 Juan, S., 'About Face', 24 Hours, January 1991, p 20.
16 Rohter, L., 'Q & A', *The New York Times*, 7 July 1987, p Y15, quoted in Juan, S., 'You Can't Tickle Yourself', *The Sydney Morning Herald*,

9 January 1991, p 12.

17 Rosenthal, E., 'Q & A', *The New York Times*, 7 November 1989, p B7, quoted in Juan, S., 'Santa's Big Belly Comes in for Some Serious Prodding', *The Sydney Morning Herald*, 21 December 1989, p 10.

18 Weiss, R., 'Travel Can Be Sickening; Now Scientists Know Why', *The New York Times*, 28 April 1992, p B5.

19 Schmeck, H., 'Q & A', *The New York Times*, 11 August 1987, p Y23.

20 Juan, S., 'Avoiding Motion Sickness', *Australian DR Weekly*, 30 October 1987, p 28.

21 Weiss, R., op. cit.

22 Associated Press, 'Motion Sickness', *The San Jose Mercury News*, 14 September 1988, p B3.

23 Livingston, K., 'Trying to Stop Motion Sickness', *The San Francisco Chronicle*, 7 December 1992, p C13, quoted in Juan, S., 'Stop the Car, Now!', *The Sydney Morning Herald*, 3 August 1989, p 10.

24 Angier, N., 'Q & A', *The New York Times*, 5 July 1988, p B11.

25 Wood, C., 'ABO Blood Groups Related to the Selection of Human Hosts by Yellow Fever Vector', *Human Biology*, 1976, 48: 337-349.

26 Jorgensen, G., 'A Contribution to the Hypothesis of a "Little More Fitness" of Blood Group O', *Journal of Human Evolution*, 1977, 6: 741-754.

27 Beardmore, J. & Karimi-Booshehri, F., 'ABO Genes Are Differentially Distributed in Socio-Economic Groups in England', *Nature*, 1983, 303: 522-524.

28 Nagorka, J., 'Genetic Key To Blood Types Found', *The Dallas Morning News*, 15 July 1990, p D 14, quoted in Juan, S., 'A Bloody Mystery Remains Unsolved', *The Sydney Morning Herald*, 6 July 1994, p 11.

10

Endings

1 AFP, 'Doc Died Happy', *The Sydney Morning Herald*, 23 May 1989, p 10.

2 Gould, G. & Pyle, W., *Anomalies and Curiosities of Medicine*, Philadelphia, W.B. Saunders, 1897, p 523.

3 Corliss, W., *Biological Anomalies: Humans I*, Glen Arm, Maryland, The Sourcebook Project, 1992, p 78.

4 Michell, J. & Rickard, R., *Phenomena: A Book of Wonders*, New York, Pantheon Books, 1977, p 24.

5 Cooper, D. & Cooper, R., 'On the Luminosity of the Human Subject After Death, With Remarks and Details of Experiments Made With a View of Determining the Nature of the Fuel', *Philosophical Magazine*, 1838, 3: 12: 420–426.

6 Landy, H., Weiner, S., Corson, S., Batzer, F. & Bolognese, R., 'The "Vanishing Twin": Ultrasonographic Assessment of Foetal Disappearance in the First Trimester', *American Journal of Obstetrics and Gynecology*, 1986, 155: 14–19.

7 Juan, S., *Only Human*, Sydney, Random House Australia, 1990, pp 103–104.

8 Brown-Sequard, C., 'Attitudes After Death', *Knowledge*, 1884, 6: 115–118.

9 Rogo, D., *The Return From Silence*, Wellingborough, Aquarian Press, 1989, p 24.

10 Moody, R., *Life After Life*, New York, Mockingbird Books, 1975.

11 Morse, M., *Transformed By the Light*, New York, Villard Books, 1992.

12 Ring, K., *Life After Death*, New York, Quill, 1982, and Ring, K., *The Omega Project*, New York, William Morrow, 1992.

13 Mauro, J., 'Bright Lights, Big Mystery', *Psychology Today*, July/August 1992, pp 54–57, 80–82, p 82, quoted in Juan, S., 'Memories Brought Back From Death', *The Times On Sunday*, 1 March 1987, p 32.

14 Stack, S. & Lester, D. 'Born Under a Bad Sign? Astrological Sign and Suicide Ideation', *Perceptual and Motor Skills*, 1988, 66: 461–462, p 461.

15 Juan, S., 'A Tendency Towards Suicide May Just Be a Bad Sign From the Sun', *The Sydney Morning Herald*, 23 November 1989, p 17.

16 Reuters, 'Boy Keeps Bursting Into Flames', *The Sydney Morning Herald*,

2 May 1990, p 19.

17 Randles, J. & Hough, P., *Death By Supernatural Causes*, London, Grafton, 1989.

18 Knight, B., 'Rainy-Day Read', *New Scientist*, 28 January 1989, p 74.

19 Juan, S., 'Combustion Theory Goes Up in Smoke', *The Sydney Morning Herald*, 8 March 1989, p 17.

20 Lieber, A., 'Human Aggression and the Lunar Synodic Cycle', *Journal of Clinical Psychiatry*, 1978, 39: 385-393.

21 Lester, D., 'Temporal Variation in Suicide and Homicide', *American Journal of Epidemiology*, 1979, 109: 517-520.

22 Little, G., Bowers, R. & Little, L., 'Geophysical Variables and Behaviour', *Perceptual and Motor Skills*, 1987, 64: 1212.

23 Coates, W., Jehle, D. & Cottington, E., 'Trauma and the Full Moon: A Waning Theory', *Annals of Emergency Medicine*, 1989, 18: 763-765, quoted in Juan, S., 'Does a Full Moon Turn Us Into Lovers Or Lunatics?', *The Sydney Morning Herald*, 22 February 1990, p 15.

24 Lipa, B., Sturrock, P. & Rogot, F., 'Search For Correlation Between Geomagnetic Disturbances and Mortality', *Nature*, 1976, 259: 302-304.

25 Tatsanaviavat, P., Chiravatkul, A., Klungboonkrong, V., Chaisiri, S., Jarerntanyaruk, L., Munger, R. & Saowa-kontha, S., 'Sudden and Unexplained Deaths in Sleep (Laitai) of Young Men in Rural Northeastern Thailand', *International Journal of Epidemiology*, 1992, 21: 904-910.

26 Lown, B., DeSilva, R. & Lenson, R., 'Roles of Psychologic Stress and Autonomic Nervous System Changes in Provocation of Ventricular Premature Complexes', *American Journal of Cardiology*, Afterword, 1978, 41: 979-985.

27 Juan, S., 'Asian Men Victims of Sudden Death', *The Sydney Morning Herald*, 21 January 1993, p 12.

28 Juan, S., 'Exhuming the Truth About Zombies Can Be Sickening', *The Sydney Morning Herald*, 16 May 1991, p 15.

29 Juan, S., 'An Afterlife Mint in Mummification', *The Sydney Morning Herald*, 12 September 1991, p 12.

Index

About the type

This book is set in Fresco, a typeface designed by Fred
Smeijers and released in 2003 by OurType. Fresco is a
large type family which includes serif and matching sans
serif variants, plus the playful 'casual' version used here
for the questions and headings. Smeijers describes its
character as 'a refreshment of traditional and conventional issues: it is definitely a contemporary typeface,
shamelessly embracing all the good given by tradition.'

Fred Smeijers (b. 1961) is one of the leading members
of the new wave of type designers working in The
Netherlands. Among his typefaces are Quadraat, Renard,
Arnhem and Sansa. In 2000 he was awarded the Gerrit
Noordzij Prize. His books *Counterpunch* (1996) and *Type
Now* (2003), both published by Hyphen Press, are among
the clearest writings on contemporary type anywhere.